D0910374

Assessing Irregular Warfare

A Framework for Intelligence Analysis

Eric V. Larson, Derek Eaton,
Brian Nichiporuk, Thomas S. Szayna

Prepared for the United States Army
Approved for public release; distribution unlimited

ARROYO CENTER

The research described in this report was sponsored by the United States Army under Contract No. DASW01-01-C-0003.

Library of Congress Cataloging-in-Publication Data

Assessing irregular warfare : a framework for intelligence analysis / Eric V. Larson ... [et al.].
   p. cm.
   Includes bibliographical references.
   ISBN 978-0-8330-4322-1 (pbk. : alk. paper)
   1. Military intelligence—United States. 2. Asymmetric warfare. I. Larson, Eric V. (Eric Victor), 1957–

UB251.U5A77 2008
355.3'432—dc22

                                                                    2008004727

The RAND Corporation is a nonprofit research organization providing objective analysis and effective solutions that address the challenges facing the public and private sectors around the world. RAND's publications do not necessarily reflect the opinions of its research clients and sponsors.

RAND® is a registered trademark.

Published 2008 by the RAND Corporation
1776 Main Street, P.O. Box 2138, Santa Monica, CA 90407-2138
1200 South Hayes Street, Arlington, VA 22202-5050
4570 Fifth Avenue, Suite 600, Pittsburgh, PA 15213-2665
RAND URL: http://www.rand.org/
To order RAND documents or to obtain additional information, contact
Distribution Services: Telephone: (310) 451-7002;
Fax: (310) 451-6915; Email: order@rand.org

Preface

This monograph documents the results of a study titled "Planning Intelligence Support to Irregular Warfare." The aim of the study was to assist the National Ground Intelligence Center (NGIC) in better understanding the intelligence analytic requirements of irregular warfare (IW) by providing an analytic framework for IW on which to base an educational and training curriculum that would enhance the capabilities NGIC analysts use to assess IW situations.

The results described should be of interest to intelligence analysts and managers in the intelligence community who are wrestling with the innumerable conceptual, collection, and analytic challenges presented by contemporary IW environments. Additionally, these results may be of interest to scholarly audiences involved in developing new analytic methodologies and tools that might be employed in IW analysis.

This research was sponsored by the NGIC, a major subordinate command of the U.S. Army's Intelligence and Security Command, and conducted within RAND Arroyo Center's Strategy, Doctrine, and Resources Program. RAND Arroyo Center, part of the RAND Corporation, is a federally funded research and development center sponsored by the United States Army.

For comments or further information about this monograph, please contact Thomas Szayna (telephone 310-393-0411, extension 7758; e-mail Thomas_Szayna@rand.org) or Eric Larson (telephone 310-393-0411, extension 7467; email larson@rand.org).

The Project Unique Identification Code (PUIC) for the project that produced this document is NGIC-06001.

For more information on RAND Arroyo Center, contact the Director of Operations (telephone 310-393-0411, extension 6419; FAX 310-451-6952; email Marcy_Agmon@rand.org), or visit Arroyo's Web site at http://www.rand.org/ard/.

Contents

Figures

Tables

Summary

The aim of this study was to assist the Department of the Army's National Ground Intelligence Center (NGIC) in better understanding the intelligence analytic requirements of irregular warfare (IW). To do this, we were to develop an analytic framework for IW that could be used as the basis for an educational and training curriculum that would enhance NGIC analysts' capabilities for assessing IW situations.

In December 2006, after considering a number of alternative definitions for *irregular warfare* and acknowledging the many conceptual and other challenges associated with trying to define this term with precision, the Office of the Secretary of Defense and the Joint Chiefs of Staff approved the following definition:

> A violent struggle among state and non-state actors for legitimacy and influence over the relevant population.

Definitions aside, large numbers of academic, doctrinal, and other publications stress that the outcomes of IW situations depend on both the level of one's understanding of the population and the deftness with which non-military and indirect means are employed to influence and build legitimacy. Accordingly, the study team's principal efforts were devoted to developing an analytic framework for understanding IW situations, whether population-centric (such as counterinsurgency) or counterterrorism, that focused on "irregular features" of the operating environment—that is, the central environmental and operational variables whose interplay determines the overall trajectory of an irregular conflict toward either success or failure.

The central idea of the framework is that it is an analytic procedure by which an analyst, beginning with a generic and broad understanding of a conflict and its environment and then engaging in successively more-focused and more-detailed analyses of selective topics, can develop an understanding of the conflict and can uncover the key drivers behind such phenomena as orientation toward principal protagonists in the conflict, mobilization, and recruitment, and choice of political bargaining or violence. Put another way, the framework allows the analyst to efficiently decompose and understand the features of IW situations—whether they are of the population-centric or the counterterrorism variety—by illuminating areas in which additional detailed analysis could matter and areas in which it probably will not matter. This analytic procedure involves three main activities and eight discrete steps, as shown in Figure S.1.

In the first activity, *initial assessment and data gathering*, the analyst focuses on developing background information on the IW operating environment. Step 1 provides the necessary background and

Figure S.1
Analytic Framework for IW Analysis

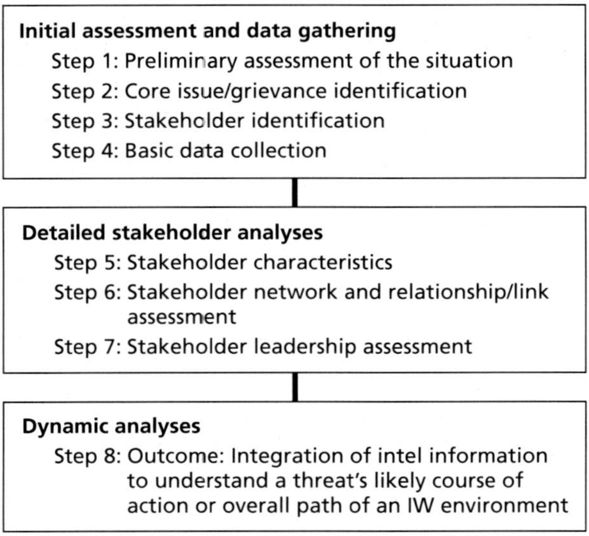

Initial assessment and data gathering
  Step 1: Preliminary assessment of the situation
  Step 2: Core issue/grievance identification
  Step 3: Stakeholder identification
  Step 4: Basic data collection

Detailed stakeholder analyses
  Step 5: Stakeholder characteristics
  Step 6: Stakeholder network and relationship/link
    assessment
  Step 7: Stakeholder leadership assessment

Dynamic analyses
  Step 8: Outcome: Integration of intel information
    to understand a threat's likely course of
    action or overall path of an IW environment

RAND *MG668-S.1*

context for understanding the situation; step 2 identifies core issues or grievances that need to be mitigated or resolved if the sources of conflict are to be eliminated; step 3 identifies key stakeholders who will seek to influence the outcome of the situation; step 4 focuses on compiling demographic, economic, attitude, and other quantitative data.

In the second activity, *detailed stakeholder analyses*, the analyst conducts a more intensive analysis of each stakeholder. Step 5 is an assessment of each stakeholder's aims, characteristics, and capabilities, both military and non-military; step 6 is an analysis of leaders, factions, and/or networks within each stakeholder group, as well as connections to other stakeholder groups and their leaders; step 7 is an analysis of key leaders identified in step 6.

In the third activity, *dynamic analyses*, the aim is to make sense of the data and insights collected in the previous steps. Much like what occurs in the intelligence preparation of the battlefield (IPB) process, step 8 consists of integrating intelligence information to determine various stakeholder groups' likely courses of action (COAs) and develop an understanding of the situation's possible trajectory. Dynamic analyses can include a wide variety of activities—for instance, trend analyses of significant activities data, content analysis of leadership statements and media, and analysis of attitude data from public opinion surveys, as well as the use of models and other diagnostic or predictive tools.

Although most of our effort focused on population-centric IW situations, available doctrine for intelligence analysis of IW suggests few distinctions between the intelligence analytic requirements of counterinsurgency and those of counterterrorism. Likewise, our analytic framework can be used for intelligence analysis in support of either population-centric IW situations, such as counterinsurgency, or counterterrorism. For example, at the tactical and operational level, terrorist organizations can be viewed as a unique class of stakeholder group or network that can be subjected to link analyses, assessments of military and non-military capabilities, leadership analyses, and other analytic activities envisioned in our framework. And when such groups are viewed as a global counterinsurgency involving transnational jihadist networks, such as the Al Qaeda organization, the distinctions between counterinsurgency and counterterrorism diminish further.

Our review of military doctrine related to IPB and IW intelligence analysis also suggests that our framework is generally compatible with the IPB process and with specific approaches, techniques, and tools advocated in existing doctrine. Incorporation of our framework—in part or in its totality—into existing intelligence analytic processes and educational and training curricula should therefore be relatively easy. In consequence, our analytic framework might best be viewed not as an alternative or competitor to IPB, but as providing an efficient analytic protocol for IW IPB analysis, one able to accent irregular features at the strategic and operational levels that are important determinants of IW outcomes.

Acknowledgments

We are grateful to Ernie Gurany, Chief Analyst at the National Ground Intelligence Center, for sponsoring the research that led to this report. We are also grateful to John S. White for monitoring the study's progress and for supporting and helping the research staff.

Two military fellows at RAND, LCDR Anthony Butera and LTC Mark Davis, provided helpful feedback on the project's progress. Brian Shannon provided research assistance.

Finally, we thank RAND colleague Gregory F. Treverton and Conrad C. Crane of the U.S. Army War College for their helpful reviews of an early draft of this monograph.

Abbreviations

ASCOPE	area, structures, capabilities, organizations, people, and events
C3	command, control, and communications
CADD	Combined Arms Doctrine Directorate
CIST	countering ideological support for terrorism
CMO	civil-military operations
COA	course of action
COE	common (or contemporary) operating environment
DoD	Department of Defense
FM	Field Manual
FMI	Field Manual–Interim
FSTC	Foreign Science and Technology Center
GMI	general military intelligence
IED	improvised explosive device
INSCOM	Intelligence and Security Command
IO	information operations
IPB	intelligence preparation of the battlefield
ITAC	Intelligence and Threat Analysis Center
IW	irregular warfare
JCA	Joint Capability Area
JOC	Joint Operating Concept
JP	Joint Publication
LLO	logical line of operation

METT-TC	mission, enemy, terrain, troops available, time, and civilian considerations
NGIC	National Ground Intelligence Center
NMSP-WOT	*National Military Strategic Plan for the War on Terrorism*
OOB	order of battle
OPCON	operational control
POM	Program Objectives Memorandum
PRIO	International Peace Research Institute, Oslo
QDR	Quadrennial Defense Review
S&TI	scientific and technical intelligence
SI	support to insurgency
SIPRI	Stockholm International Peace Research Institute
SOCOM	Special Operations Command
SOF	Special Operations Forces
SSTRO	stabilization, security, transition, and reconstruction operations
TTP	tactics, techniques, and procedures
UW	unconventional warfare
WMD/E	weapons of mass destruction/effects

CHAPTER ONE

Introduction

Background to the Study

The sponsor of our study, the National Ground Intelligence Center (NGIC), is the primary producer of ground forces intelligence in the Department of Defense (DoD).[1] NGIC was created in March 1995, when the U.S. Army Foreign Science and Technology Center (FSTC) and the U.S. Army Intelligence and Threat Analysis Center (ITAC) were merged to form a Center of Excellence devoted to providing ground-component intelligence-production support to national and departmental intelligence consumers.[2] Headquarters, U.S. Army Intelligence and Security Command (INSCOM), exercises direct operational control (OPCON) over NGIC, which is a major subordinate command of INSCOM. NGIC's mission statement is

> [T]o produce all-source integrated intelligence on foreign ground forces and support combat technologies to ensure that U.S. forces and other decision makers will always have a decisive edge on any battlefield.[3]

And its institutional vision is

---

[1] See Robert O'Connell and John S. White, "NGIC: Penetrating the Fog of War," *Military Intelligence Professional Bulletin*, April–June 2002, pp. 14–18.

[2] See DoD, "Memorandum for Correspondents," Memorandum No. 046-M, March 2, 1995.

[3] National Ground Intelligence Center, Web site home page, December 2006.

> [To be the] Premier Intelligence analysis organization in DoD
> . . . [f]rom analytic products that ensure U.S. forces and their
> allies will always have a decisive edge in equipment, organization,
> and training on any future battlefield . . . [t]o on-the-spot intel-
> ligence for the fight . . . [t]o providing information that affects
> policy decisions at all levels . . . [i]n an organizational environ-
> ment of trust, respect, and communications dedicated to selfless
> service for the nation.[4]

NGIC produces multi-source intelligence products that include scientific and technical intelligence (S&TI) and general military intelligence
(GMI) on foreign ground forces in support of combatant commanders,
force and material developers, the Department of the Army, DoD, and
other national-level decisionmakers. Historically, NGIC has produced
and maintained intelligence on foreign scientific developments, ground
force weapons systems, and associated technologies.[5] NGIC aspires to
be the Center of Excellence for ground force irregular warfare (IW)
intelligence production.[6]

---

[4]   National Ground Intelligence Center, Web site home page, "About" section, December
2006.

[5]   NGIC analysis includes but is not limited to military communications electronics systems;
types of aircraft used by foreign ground forces; nuclear, biological, and chemical (NBC) systems; and basic research in civilian technologies with possible military applications. Headquarters, Department of the Army, *Intelligence*, Field Manual (FM) 2-0, Washington, D.C.,
May 2004, p. 10-2.

[6]   According to reports in the media, the classified 2006 Quadrennial Defense Review
(QDR) execution roadmap on IW called for the establishment of a Center of Excellence
for IW. See Sebastian Sprenger, "DOD, State Dept. Eye Joint 'Hub,'" *Inside the Pentagon*,
November 16, 2006. The role of the center would be to "coordinate IW research, educa-
tion, training, doctrine, and lessons learned" (Joint Chiefs of Staff, "Irregular Warfare (IW)
Execution Roadmap," unclassified briefing, undated, slides 9–13). The Marine Corps has a
Small Wars Center of Excellence at Quantico, Virginia, and the Air Force has stood up an
IW Center of Excellence at Nellis Air Force Base that aims to "give our foreign and potential
coalition partners a one-stop shop for all integration issues with the Air Force." (Quotation
from Secretary of the Air Force Michael W. Wynne, "State of the Force," remarks to Air
Force Association's Air and Space Conference and Technology Exposition 2006, Washington, D.C., September 25, 2006. Information on the Small Wars Center of Excellence can be
found at the U.S. Marine Corps Small Wars web page, 2007.

This aspiration—and the impetus for the study—derives from the recent emergence and growing importance of IW in DoD:

> [W]e must display a mastery of irregular warfare comparable to that which we possess in conventional combat. . . . [I]mproving the U.S. Armed Forces' proficiency in irregular warfare is the Defense Department's top priority.[7]

Study Aims and Analytic Approach

Following its designation as a Center of Excellence for IW intelligence production, NGIC asked RAND to provide assistance in developing an education and training curriculum for improving the capabilities available to NGIC analysts for IW-related intelligence analyses.

In consultation with the sponsor, we divided the problem into two phases. The first focused on identifying the intelligence and analytic requirements associated with IW and developing a framework for intelligence analysis of IW operating environments that subsequently could be translated into an education and training curriculum. The goal of the second phase was to translate this framework into a more detailed education and training curriculum for NGIC. This monograph documents the results of the first phase of the overall effort.

Figure 1.1 describes the approach we took in identifying IW intelligence and analytic requirements. As the figure shows, the study team took three separate passes at the problem.

The team's first pass involved a review of extant Army and other U.S. military doctrine to understand what intelligence and analytic requirements of IW already had been identified. The doctrinal review included a review of mission-oriented doctrine for IW's constituent

---

7   See DoD, *National Defense Strategy*, Washington, D.C., June 2008, pp. 4, 13. Secretary of Defense Robert Gates has identified the "long war" against violent extremism as the nation's top priority over coming decades. See Josh White, "Gates Sees Terrorism Remaining Enemy No. 1; New Defense Strategy Shifts Focus from Conventional Warfare," *The Washington Post*, July 31, 2008, p. A1.

Figure 1.1
Analytic Approach for Identifying IW Intelligence and Analytic Requirements

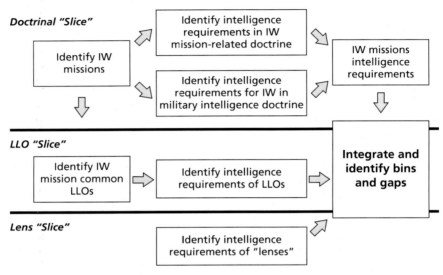

RAND *MG668-1.1*

missions (more on this later), as well as a review of military intelligence doctrine to see what it had to say about IW.

The second pass took a different approach. The team began by identifying common logical lines of operation (LLOs) for IW's constituent missions that had been identified in U.S. military doctrine and other publications; it then held brainstorming sessions to identify the intelligence and analytic requirements associated with each LLO.[8]

---

[8]   Lines of operation "define the directional orientation of the force in time and space in relation to the enemy. They connect the force with its base of operations and its objectives" (DoD, *Department of Defense Dictionary of Military and Associated Terms*, Joint Publication (JP) 1-02, Washington, D.C., April 12, 2001 (as amended through April 14, 2006), p. 310). In contrast, logical lines of operation, or LLOs, "define the operational design when positional reference to an adversary has little relevance. . . . Operations designed using LLOs typically consist of an extended, event-driven time line. This time line combines the complementary, long-range effects of civil-military operations as well as the cyclic, short-range events characteristic of combat operations" (Headquarters, Department of the Army, *The Operations Process*, Field Manual–Interim (FMI) 5-0.1, Washington, D.C., March 2006,

The third pass, which was based on insights from past RAND work and a review of the academic literature,[9] viewed the IW environment through different methodological "lenses," including expected utility modeling, social network analysis, media content or communications analysis, public opinion analysis, and major theories related to IW, mobilization, and other relevant phenomena.

These parallel efforts led to lists of IW intelligence and analytic requirements that we compiled and taxonomically organized. To assess the comprehensiveness and completeness of these lists of requirements, we then cross-checked them with area study outlines, educational curricula, and military intelligence, academic, and other syllabi that had been developed for the study of IW, as well as with other materials.

In developing a framework for IW intelligence analysis, the study team aimed to identify those features of the IW environment that best captured the inherently dynamic and changing character of IW situations, including mobilization, escalation, coalition formation, bargaining, and influence. Ultimately, this led to a logically related set of analytic tasks that, taken together, are highly likely to lead to complete and comprehensive analyses of any given IW environment.

Organization of This Monograph

Chapter Two of this monograph evaluates IW through a review of recent DoD efforts to define IW; Chapter Three reviews the analytic requirements of IW and presents the analytic framework the study team developed for assessing IW situations; Chapter Four provides conclusions. Appendix A is a review of official policy and strategy documents

---

pp. A-6 and A-7. The LLOs we considered were combat operations, training and employing host nation security forces, governance, essential services, economic development, and strategic communications/information operations.

[9]   Eric V. Larson et al., *Foundations of Effective Influence Operations*, MG-654-A, Santa Monica, Calif.: RAND Corporation, forthcoming; and Eric V. Larson et al., *Understanding Commanders' Information Needs for Influence Operations*, MG-656-A, Santa Monica, Calif.: RAND Corporation, forthcoming.

related to IW, and Appendix B lists doctrinal publications identified as addressing the intelligence analytic requirements of IW.

CHAPTER TWO

Defining Irregular Warfare

Historical U.S. experience with internal conflicts around the world provides ample testimony to the challenges of conducting successful military operations in environments where military and political factors are tightly interwoven—consider, for example, the Philippines and China at the turn of the 20th century, Russia after World War I, Central America and the Caribbean in the 1920s and 1930s, the Chinese civil war after World War II, Vietnam in the 1960s, Lebanon in the 1980s, Somalia in the 1990s, and Afghanistan and Iraq in the present decade.[1] Intrastate conflicts are the most prevalent form of warfare in the world.[2] Thus, even if the United States has been more selective about direct military involvement in such conflicts than this list suggests,[3] U.S. participation in future IW operations has been and is

---

[1] See, for example, Frank G. Hoffman, "Small Wars Revisited: The United States and Nontraditional Wars," *The Journal of Strategic Studies*, Vol. 28, No. 6, December 2005, pp. 913–940.

[2] According to researchers at the Uppsala Conflict Data Program (probably the foremost organization monitoring armed conflict), of the 121 armed conflicts logged from 1989 to 2005, 90 were intrastate, 24 were internationalized intrastate, and seven were interstate. In 2005, 25 intrastate armed conflicts, six intrastate armed conflicts in which foreign governments supported one side, and no interstate conflicts occurred (Lotta Harbom, Stina Hogbladh, and Peter Wallensteen, "Armed Conflict and Peace Agreements," *Journal of Peace Research*, Vol. 43, No. 5, 2006, Table II, p. 618).

[3] In 2005, 31 intrastate and internationalized intrastate conflicts took place, with U.S. forces directly involved in three of them: Iraq, Afghanistan, and the global campaign against the Al Qaeda organization (Harbom, Hogbladh, and Wallensteen, "Armed Conflict and Peace Agreements," 2006, Appendix II, pp. 627–630). Obviously not included in these num-

7

likely to remain—barring a fundamental redefinition of U.S. interests—a persistent feature of U.S. defense policy.[4]

IW's salience to the defense community, moreover, has recently increased. This is largely a result of the strategic imperative of countering the threat posed by the Al Qaeda organization's transnational jihadist movement, and the range of specific challenges the United States has encountered in the Afghan and Iraqi insurgencies, which recently led to a high degree of policy- and strategy-level attention to the requirements of IW.[5]

In this chapter, we review alternative definitions of IW that are used within DoD, enumerate the types of operations generally conceived as constituting IW, and discuss the principal campaign-level tasks that underwrite IW operations.

A Review of Recent Efforts to Define Irregular Warfare

Until recently, DoD had no single approved doctrinal definition of *irregular warfare*; efforts to define the term had been contentious, and the results somewhat problematic. The January 2007 draft of the IW Joint Operating Concept (JOC) acknowledged the definitional difficulties in language that was then used in the September 2007 release of the IW JOC:

> IW is a complex, "messy," and ambiguous social phenomenon that does not lend itself to clean, neat, concise, or precise definition. This JOC uses the term in two contexts. First, IW is a form of armed conflict. As such, it replaces the term "low-intensity con-

---

bers are missions involving the training of host nation forces and other specific missions involving primarily special forces.

[4]   Lawrence A. Yates, *The U.S. Military's Experience in Stability Operations, 1789–2005*, Global War on Terrorism Occasional Paper 15, Fort Leavenworth, Kan.: Combat Studies Institute Press, 2006.

[5]   Appendix A reviews discussions of IW found in recent official policy and strategy documents.

flict." Second, IW is a form of warfare. As such, it encompasses insurgency, counterinsurgency, terrorism, and counterterrorism, raising them above the perception that they are somehow a lesser form of conflict below the threshold of warfare.[6]

The difficulties of defining IW are apparent in two alternative definitions that recently competed for official status within DoD.[7] The first of these emerged from a September 2005 IW workshop hosted by the U.S. Special Operations Command (SOCOM) and the Office of the Assistant Secretary of Defense for Special Operations and Low-Intensity Conflict.[8] This definition subsequently was modified before ultimately being approved by Deputy Secretary of Defense Gordon England on April 17, 2006:[9]

---

[6] DoD, *Irregular Warfare (IW) Joint Operating Concept (JOC)*, January 2007, p. 4; and DoD, *Irregular Warfare (IW) Joint Operating Concept (JOC)*, Version 1.0, September 2007, p. 6. These two documents are, from here on, referred to as IW JOC 1/07 and IW JOC 9/07, respectively. For its part, the Air Force's doctrine document for IW, dated August 1, 2007, simply states that IW "does not easily lend itself to a concise universal definition." See United States Air Force, *Irregular Warfare*, AFDD 2-3, Washington, D.C., August 1, 2007, p. 11.

[7] A fairly comprehensive list of official efforts to define IW—and a scathing critique of IW as an organizing concept—is to be found in U.S. Joint Forces Command Joint Warfighting Center, *Irregular Warfare Special Study*, Washington, D.C., August 4, 2006. Additionally, the U.S. Marine Corps Combat Development Command's June 2006 *Tentative Manual for Countering Irregular Threats: An Updated Approach to Counterinsurgency Operations* (Quantico, Va., June 7, 2006, p. 1) states that "[t]he term irregular is used in the broad, inclusive sense to refer to all types of non-conventional methods of violence employed to counter the traditional capabilities of an opponent. Irregular threats include acts of a military, political, psychological, and economic nature, conducted by both indigenous actors and non-state actors for the purpose of eliminating or weakening the authority of a local government or influencing an outside power, and using primarily asymmetric methods. Included in this broad category are the activities of insurgents, guerrillas, terrorists, and similar irregular groups and organizations that operate in and from the numerous weakened and failed states that exist today."

[8] These origins are described in U.S. Joint Forces Command Joint Warfighting Center, *Irregular Warfare Special Study*, 2006, Enclosure L.

[9] For example, a slightly modified version of the definition that opened with "The ability to conduct warfare . . ." is contained in Joint Staff, "Proposed Joint Capability Areas Tier 1 and Supporting Tier 2 Lexicon (Mar 06 refinement effort results)," Washington, D.C., March 2006, p. 13.

> A form of warfare that has as its objective the credibility and/
> or legitimacy of the relevant political authority with the goal
> of undermining or supporting that authority. Irregular warfare
> favors indirect approaches, though it may apply the full range of
> military and other capabilities to seek asymmetric approaches, in
> order to erode an adversary's power, influence and will.

Since its approval, this definition has been widely used in a number of official DoD publications.[10]

As of late October 2006, however, another definition of IW had been introduced. It appeared in the final Coordination Draft of *Joint Warfare of the Armed Forces of the United States* (former title of JP 1) and was included in both the May 2007 final version of JP 1 and IW JOC 9/07:

> A violent struggle among state and non-state actors for legiti-
> macy and influence over the relevant populations. IW favors
> indirect and asymmetric approaches, though it may employ the
> full range of military and other capacities, in order to erode an
> adversary's power, influence, and will. It is inherently a protracted
> struggle that will test the resolve of our Nation and our strategic
> partners.[11]

It thus appears that this definition of IW has supplanted the earlier one and is the authoritative definition within DoD.

---

[10] It has been used in Joint Chiefs of Staff, "Irregular Warfare (IW) Execution Roadmap," undated; U.S. Marine Corps Combat Development Command and U.S. Special Operations Command Center for Knowledge and Futures, *Multi-Service Concept for Irregular Warfare*, Version 2.0, August 2, 2006, p. 7; and Statement of Brigadier General Otis G. Mannon, U.S. Air Force, Deputy Director, Special Operations, J-3, Joint Staff, Before the 109th Congress Committee on Armed Services, Subcommittee on Terrorism, Unconventional Threats and Capabilities, United States House of Representatives, September 27, 2006.

[11] DoD, *Doctrine for the Armed Forces of the United States*, JP 1, Washington, D.C., May 14, 2007, p. I-1; and IW JOC 9/07, p. 1. Most recently, the 2008 National Defense Strategy picks up this language, describing the war against Al Qaeda and its associates as "a prolonged irregular campaign, a violent struggle for legitimacy and influence over the population." See DoD, *National Defense Strategy*, Washington, D.C., June 2008, p. 4.

Despite the differences between these two definitions—and their failure to fully eliminate the somewhat nebulous nature of IW—they do appear to share one principal feature: a set of operating environment characteristics very different from those associated with success in conventional warfare.[12] First, the threats generally are asymmetric or irregular rather than conventional. Second, success hinges in large measure not on defeating forces but on winning the support or allegiance—or defeating the will—of populations. On this second point, both definitions emphasize that such psychological concepts as credibility, legitimacy, and will are the central focus in IW. They also emphasize such political concepts as power and influence in the competition for sympathy from, support from, and mobilization of various segments of the population, as well as a reliance on indirect and non-military rather than military approaches. Finally, both imply that the use of violence must be carefully calibrated so as to ensure that it does more harm than good in the attempt to win support from the indigenous population.[13]

Irregular Warfare Operation Types

A number of efforts have also been made to define the specific missions that make up IW. These, too, have had somewhat inconsistent results:

- Although the February 2006 *Quadrennial Defense Review* (QDR) *Report* used the term *irregular warfare* in varying ways, it explicitly called out the following as missions in the IW portfolio: counterinsurgency; unconventional warfare; stability, or stabilization,

---

[12] For a detailed analysis of the difficulties of using IW as an organizing concept for joint doctrine development, see U.S. Joint Forces Command Joint Warfighting Center, 2006.

[13] Polling in Iraq, for example, showed that support for the U.S. coalition was negatively associated with the belief that the United States was not being careful enough in avoiding civilian casualties (Eric V. Larson and Bogdan Savych, *Misfortunes of War: Press and Public Reactions to Civilian Deaths in Wartime*, MG-441-AF, Santa Monica, Calif.: RAND Corporation, 2007, pp. 200–202).

security, transition, and reconstruction operations (SSTRO); and counterterrorism.[14]

- The U.S. Army's Combined Arms Doctrine Directorate (CADD) treats IW as consisting of four distinct missions: counterinsurgency; support to insurgency; foreign internal defense; counterterrorism.[15]
- The Joint Staff's August 2006 proposed taxonomy for Joint Capability Areas (JCAs) treated IW as a Tier 2 JCA, part of the Joint Special Operations and Irregular Warfare Tier 1 JCA, and identified counterinsurgency and foreign internal defense as Tier 2 missions; but it also identified unconventional warfare, counterterrorism, psychological operations, and civil-military operations as Tier 3 Special Operations Forces (SOF) JCAs that support IW.[16]
- The August Multi-Service Concept for Irregular Warfare, while accenting offensive operations (e.g., unconventional warfare, counterterrorism), also identifies missions and activities, including counterinsurgency, foreign internal defense, support to insurgency, unconventional warfare, stability/SSTRO, counterterrorism, psychological operations, civil-military operations, information operations, and intelligence/counterintelligence.[17]
- IW JOC 9/07 identified the following missions and activities as composing IW: insurgency; counterinsurgency; unconventional warfare; terrorism (by adversaries); counterterrorism; foreign internal defense; SSTRO; strategic communications; psychological operations; information operations; civil-military operations; intelligence and counterintelligence activities; transnational

---

[14] DoD, *Quadrennial Defense Review Report*, Washington, D.C., February 2006.

[15] U.S. Army Combined Arms Doctrine Directorate, "The Continuum of Operations and Stability Operations," briefing, Ft. Leavenworth, Kan.: U.S. Army Combined Arms Center, 2006.

[16] Joint Staff, "Joint Capability Areas Taxonomy Tier 1 & Tier 2 with the Initial Draft of Joint Force Projection," briefing, post 24 August 2006 Joint Requirements Oversight Council (JROC), Washington, D.C., August 2006.

[17] U.S. Marine Corps Combat Development Command and U.S. Special Operations Command Center for Knowledge and Futures, *Multi-Service Concept for Irregular Warfare*, 2006, p. 11.

criminal activities that support or sustain adversaries' IW activities; law enforcement activities focused on countering irregular adversaries.[18]

Table 2.1 summarizes the evolution in DoD thinking about IW missions and activities. As should be clear from this table, although these sources agree, nearly uniformly, that counterinsurgency is an IW mission/activity, and also agree, to a lesser extent, that counter-

Table 2.1
Irregular Warfare Missions and Activities

Mission/Activity	DoD Source[a]				
	QDR 2/06	CADD Briefing 2006	JCA Lexicon 8/06	MSC 8/06	IW JOC 9/07
Counterinsurgency	X	X	X	X	X
Foreign internal defense		X	X	X	X
(Support to) insurgency		X		X	X
Unconventional warfare	X		(X)	X	X
Stability/SSTRO	X			X	X
Strategic communications					X
Counterterrorism	X	X	(X)	X	X
Psychological operations			(X)	X	X
Civil-military operations			(X)	X	X
Information operations				X	X
Intelligence/counter-intelligence				X	X

NOTES: (1) JOC capability areas denoted with "(X)" are Tier 3 Special Operations and Information Operations JCAs that support IW rather than being direct IW missions. (2) The MSC and IW JOC sources include terrorism and transnational criminal activities in IW but stress that these activities violate U.S. and international law and accordingly are not employed by U.S. military forces or civilian government employees.

[a] QDR 2/06 = DoD, Quadrennial Defense Review Report, 2006, pp. 4, 38; CADD Briefing 2006 = U.S. Army Combined Arms Doctrine Directorate, "The Continuum of Operations and Stability Operations," 2006, slide 4; JCA Lexicon 8/06 = Joint Staff, "Joint Capability Areas Taxonomy Tier 1 & Tier 2 with the Initial Draft of Joint Force Protection," 2006, slide 9; MSC 8/06 = U.S. Marine Corps Combat Development Command and U.S. Special Operations Command Center for Knowledge and Futures, Multi-Service Concept for Irregular Warfare, 2006, p. 11; and IW JOC 9/07 = DoD, Irregular Warfare (IW) Joint Operating Concept (JOC), September 2007, p. 10.

---

[18] IW JOC 9/07, p. 10.

terrorism, foreign internal defense, and unconventional warfare are IW missions/activities, some important differences exist.[19] The most comprehensive—and because it has now been approved, authoritative—list of activities and missions, moreover, is to be found in IW JOC 9/07.

Thus, IW includes operations that are essentially offensive in nature (e.g., counterterrorism and support to insurgency or unconventional warfare) and operations that have a mixed, or more defensive, quality to them (e.g., counterinsurgency and foreign internal defense).[20]

It should be clear from this discussion that IW operations generally can be thought of in terms of two main types: (1) what one might call *population-centric IW*, which is marked by insurgency and counterinsurgency operations that may also include other activities (e.g., foreign internal defense, SSTRO, and counterterrorism operations); and (2) *counterterrorism operations*, whether conducted in the context of a larger counterinsurgency or other campaign or conducted independent of such operations as part of SOCOM's campaign for the war on terrorism.

Irregular Warfare Common Logical Lines of Operation

We also reviewed doctrinal and other documents to see how subordinate activities of IW operations might be binned as LLOs. Army FM 3-24 describes the use of LLOs in counterinsurgency as follows:

> Commanders use LLOs to visualize, describe, and direct operations when positional reference to enemy forces has little relevance. LLOs are appropriate for synchronizing operations against enemies that hide among the populace. A plan based on LLOs unifies the efforts of joint, interagency, multinational, and HN

---

[19] In part, this may have to do with imprecision regarding which are doctrinal *missions* and which are simply *activities*. We made no effort here to resolve this imprecision, given the uncertainties about the form in which IW might emerge from the Joint Staff process.

[20] Nowhere is this clearer than in the cases of counterinsurgency and support to insurgency, which essentially are opposites—the first in support of the government; the second in support of the opposition.

[host nation] forces toward a common purpose. Each LLO represents a conceptual category along which the HN government and COIN [counterinsurgency] force commander intend to attack the insurgent strategy and establish HN government legitimacy. LLOs are closely related. Successful achievement of the end state requires careful coordination of actions undertaken along all LLOs.[21]

Table 2.2 lists a number of LLOs that our review of doctrine identified as typically associated with IW operations. As can be seen, the doctrinal sources we reviewed suggest that there is substantial agreement about combat operations, training and employment of host nation security and military forces, governance, essential services, and economic development being critical lines of operation that span IW. Some documents also suggest that strategic communications and information operations and intelligence should be included as separate lines of operation; indeed, FM 3-24, *Counterinsurgency*, takes the view that strategic communications and information operations

Table 2.2
Irregular Warfare Logical Lines of Operation

	DoD Sources[a]		
	MSC 8/06	FM 3-24 12/06	IW JOC 9/07
Combat operations	X	X	X
Host nation security forces	X	X	X
Governance	X	X	X
Essential services	X	X	X
Economic development	X	X	X
Strategic communications/ information operations	X	X	X
Intelligence			X

[a] MSC 8/06 = U.S. Marine Corps Combat Development Command and U.S. Special Operations Command Center for Knowledge and Futures, Multi-Service Concept for Irregular Warfare, 2006, p. 6; FM 3-24 12/06 = Headquarters, Department of the Army, Counterinsurgency, 2006, p. 5-3; IW JOC 9/07 = DoD, Irregular Warfare (IW) Joint Operating Concept (JOC), September 2007, p. 10.

---

[21] Headquarters, Department of the Army, *Counterinsurgency*, FM 3-24, Washington, D.C., December 2006, p. 5-3.

are the most important LLOs in counterinsurgency warfare.[22] Moreover, each line of operation may have distinct intelligence information and analysis requirements, as shown in Figure 2.1.[23]

The substantial agreement on the importance of these lines of operation in IW suggests that this might provide a basis for deriving

Figure 2.1
Intelligence Requirements for Irregular Warfare Logical Lines of Operation

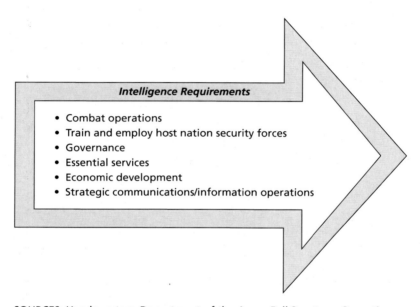

Intelligence Requirements

- Combat operations
- Train and employ host nation security forces
- Governance
- Essential services
- Economic development
- Strategic communications/information operations

SOURCES: Headquarters, Department of the Army, *Full Spectrum Operations*, initial draft, FM 3-0, June 21, 2006; Headquarters, Department of the Army, *Counterinsurgency*, 2006; IW JOC 6/07.
RAND *MG668-2.1*

---

[22] Headquarters, Department of the Army, *Counterinsurgency*, FM 3-24, 2006, especially pp. 5-8 to 5-11.

[23] FM 3-24 (Headquarters, Department of the Army, December 2006) argues that, like intelligence, each of the other lines of operation may have unique information operations and/or strategic communications requirements, and that information operations/strategic communications should therefore be viewed as a cross-cutting function.

intelligence analytic requirements that would be not only complementary, but also less susceptible to differing conceptions of the mission types making up IW.[24]

Chapter Conclusions

When one steps back from the details of a review of IW definitions, operation types, and LLOs, IW can be thought of in terms of two stylized, ideal types. The first of these, which we call *population-centric IW*, is perhaps best evidenced by typical counterinsurgency operations, such as those being conducted in Iraq and Afghanistan, where the focus is primarily on building indigenous public support (or tolerance) for U.S. aims. As described above, this type can involve a wide range of military missions, including, among others, combat and training of host nation security; but the outcome most often depends on the success of intrinsically political efforts to reach a stable political equilibrium underwritten by improvements to personal security for the population, restoration of essential services, and economic development and good governance.[25] In this ideal type, the weight of effort is focused less on military than on political, psychological, informational, and related efforts, less on defeating enemy forces than on persuading those who can be persuaded to support the U.S.-supported aims and government.

As described in the February 2006 *Quadrennial Defense Review Report*, the second ideal type is IW against "dispersed, global terrorist networks that exploit Islam to advance radical political aims."[26] This type focuses on the Al Qaeda organization umbrella of ideologically connected, cellular-structured groups; it targets specific individuals or

---

[24] It also is worth noting that the relative importance of these LLOs seems likely to vary across specific IW mission types.

[25] "Tactical and Operational competence in conventional warfighting does not necessarily guarantee tactical, operational, or strategic success in operations and activities associated with IW" (Joint Chiefs of Staff, "Irregular Warfare (IW) Execution Roadmap," undated).

[26] DoD, *Quadrennial Defense Review Report*, 2006.

small cells widely dispersed across the globe and requires an exquisite level of precision and timeliness in intelligence, targeting, and striking capability. This form of IW is highly tactical and technical in nature and generally does not rely on general-purpose forces. Instead, the principal application of military power consists of direct action by small numbers of SOF and, presumably, precision strikes by manned or unmanned aircraft.[27] It also can be prosecuted by non-military partners, including law enforcement or paramilitary direct action.

To summarize, our analysis led us to several key points. First, efforts to specify intelligence analytic needs for IW will need to consider a wide range of IW situations and "irregular features" not typically taken into account in intelligence analysis for conventional operations. Second, policy and strategy documents suggest that one can crudely divide IW into what might be thought of as two types: population-centric IW operations (such as counterinsurgency and support to insurgency) and counterterrorism operations, whether conducted within or independent of a counterinsurgency or similarly large IW operation. Third, IW remains a somewhat nebulous concept, and the defense community has had great difficulty defining and operationalizing IW in a precise and generally agreed upon manner. However, despite some inconsistency in the treatment of IW, there seems to be less disagreement about the sorts of features that one must understand to be successful in IW and the common LLOs that underwrite IW.

---

[27] We include in precision strikes the MQ-9 Reaper (formerly the RQ-1 Predator B) unmanned combat aerial vehicle, ground attack aircraft, long-range bombers, and cruise missiles. Action by civilian law enforcement or paramilitary capabilities also could be used to capture or incapacitate terrorists. We also note requirements for combating ideological support for terrorism (CIST)—the larger "war of ideas" that aims to reduce support among Muslims for extremist positions. Although CIST may be a critically important military activity in a ground commander's area of operations, we view the global campaign generally as more of a civilian than a military responsibility.

CHAPTER THREE

A Framework for Assessing Irregular Warfare

The preceding chapter discussed the possibility of viewing the IW operations of greatest policy interest as two ideal main types:

- *Population-centric IW operations.* These are characterized by counterinsurgency, foreign internal defense, and large-scale SSTRO campaigns of the kind being waged in Iraq; their success depends on some measure of security being established and a preponderance of the population being mobilized in support of U.S. aims.[1]
- *Counterterrorism operations.* These run the gamut from tactically precise direct action or raids in a larger, geographically focused IW (e.g., counterinsurgency) campaign, to the type of campaign being waged against the Al Qaeda organization, a globally dispersed network of ideologically committed jihadists cum terrorists.

In this chapter, we consider the intelligence analytic requirements of each of these two types of IW operations.

---

[1] It is worth noting that although stability operations and foreign internal defense may be performed separately from counterinsurgency, if counterinsurgency operations are under way, these two types of operations will also be under way.

Population-Centric Irregular Warfare Operations

Whereas the success of conventional warfare depends primarily on military factors, success in IW depends primarily on a wide range of *irregular features* of the operating environment—features less important in or entirely absent from conventional warfare.[2] As IW JOC 1/07 states:

> What makes IW different is the focus of its operations—a relevant population—and its strategic purpose—to gain or maintain control or influence over, and the support of that relevant population. In other words, the focus is on the legitimacy of a political authority to control or influence a relevant population.[3]

> To achieve this understanding [of the IW operating environment], the Intelligence Community will establish persistent long-duration intelligence networks that focus on the population, governments, traditional political authorities, and security forces at the national and sub-national levels in all priority countries. The joint force will leverage these networks by linking them to operational support networks of anthropologists and other social scientists with relevant expertise in the cultures and societies of the various clans, tribes, and countries involved. Where civilian expertise in the social sciences is not available, DoD will provide its own experts. Reachback to academia is useful, but not a fail-safe in extended operational environments.[4]

As should be clear from the review of IW definitions in Chapter Two, a focus on these sorts of irregular features appears to offer a more profitable approach for understanding the analytic requirements of IW than attempting to smooth out the definitional and other difficulties.

The study team's efforts concentrated primarily on developing an analytic framework for understanding population-centric IW opera-

---

[2]   We are indebted to RAND colleague Jim Quinlivan for this observation.

[3]   IW JOC 1/07, p. 5.

[4]   IW JOC 1/07, pp. 21–22.

tions that focus on the central environmental and operational variables, whose interplay determines the overall trajectory of an irregular war toward success or failure for the United States.[5]

In constructing this framework, the team aimed to provide a simple, top-down procedure for intelligence analyses of IW that would

- highlight, through a number of complementary analytic passes, the key features that drive IW situations, rather than simply compile lists
- synthesize disparate literatures (doctrine, academic) to identify alternative lenses, analytic techniques, and tools that can be employed in IW analysis
- address unique military features of IW but also focus on the political and other non-military features at the heart of IW, including the shifting sympathies and affiliations of different groups and their mobilization to political activity and the use of violence.

Put another way, the framework was designed to enable analysts to "peel the onion" and thereby uncover critical characteristics of any given IW operating environment.

The central idea of the framework is that it is an analytic procedure by which an analyst, beginning with a generic and broad understanding of a conflict and its environment and then engaging in successively more focused and detailed analyses of selective topics, can develop an understanding of the conflict and uncover the key drivers behind such phenomena as orientation toward the principal protagonists in the conflict, mobilization and recruitment, and choice of political bargaining or violence. In other words, the framework can be used to efficiently decompose and understand the features of IW situations—whether of the population-centric IW type or the counterterrorism variety—by illuminating areas in which additional detailed analysis could matter and where it probably will not make a difference.

---

[5] As described in Chapter One, the framework was the result of a number of separate analytic efforts, including a detailed doctrinal review, an assessment of the suitability of various methodological tools and models, and brainstorming sessions.

Figure 3.1 depicts this procedure. As can be seen, the procedure involves three main activities: initial analysis and data gathering, detailed stakeholder analyses, and dynamic analyses, which together make up a total of eight discrete steps. These activities and steps are described next.

Initial Assessment and Data Gathering

As shown, this activity consists of four steps, beginning with the analyst focusing on gaining an overview of the origins and history of the conflict; what various classified and unclassified secondary analyses have to say about the key political, socioeconomic, and other drivers of the conflict; and the key fault lines or other structural characteristics of the conflict (e.g., the nature of the coalitions supporting the government and its challengers).[6]

In the second step, the analyst explores in greater detail the core grievances underlying the conflict and the key proximate issues currently in contention. Among the sorts of questions of interest in this step are, What issues or grievances are being exploited to mobilize different groups? Which issues or grievances just beneath the surface of the conflict are really driving various parties to the conflict? Have these issues or grievances changed over time?

Closely following on the analysis of issues and grievances is the third step, in which the analyst identifies, in a relatively comprehensive fashion, the key stakeholders that have grievances or otherwise are likely to seek to influence the outcome of the conflict through various

---

[6] In addition to various classified products of the intelligence community, many unclassified sources provide trenchant analyses of conflicts that are under way—for example, International Crisis Group assessments (online at http://www.crisisgroup.org/home/index.cfm); Jane's World Insurgency and Terrorism (online at http://jwit.janes.com/public/jwit/index.shtml) and Sentinel Security Assessments (online at http://sentinel.janes.com/public/sentinel/index.shtml); the Stockholm International Peace Research Institute (SIPRI) yearbooks (Oxford University Press: Oxford; most recent year available at the time of writing, 2006, online at http://yearbook2006.sipri.org/) and the "UCDP/PRIO [Uppsala Conflict Data Program/International Peace Research Institute, Oslo] Armed Conflict Dataset" (online at http://www.prio.no/CSCW/Datasets/Armed-Conflict/UCDP-PRIO/); International Institute for Strategic Studies (IISS) strategic surveys (online at http://www.iiss.org/); and the Economist Intelligence Unit (online at http://www.eiu.com/).

Figure 3.1
IW Assessment Framework

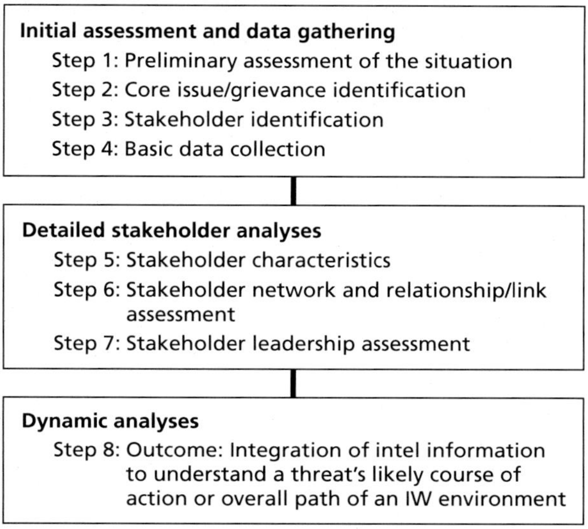

Initial assessment and data gathering
    Step 1: Preliminary assessment of the situation
    Step 2: Core issue/grievance identification
    Step 3: Stakeholder identification
    Step 4: Basic data collection

Detailed stakeholder analyses
    Step 5: Stakeholder characteristics
    Step 6: Stakeholder network and relationship/link
        assessment
    Step 7: Stakeholder leadership assessment

Dynamic analyses
    Step 8: Outcome: Integration of intel information
        to understand a threat's likely course of
        action or overall path of an IW environment

RAND *MG668-3.1*

means. This effort involves identifying major political, demographic, social, military, paramilitary, terrorist, and other groups or factions seeking or that may seek to influence the outcome. This entails looking at domestic groups, factions, movements, and other stakeholders, as well as at international and transnational institutions, groups, and actors, and states that are allies or adversaries.[7]

In the fourth step, which can be undertaken in parallel with and cued by the results of the other steps, the analyst compiles basic demographic, economic, and other quantitative data that relate to the drivers and fault lines identified in the earlier steps.[8] This effort includes col-

---

[7] In addition to various classified and unclassified intelligence products from the Central Intelligence Agency (CIA), Defense Intelligence Agency (DIA), and Open Source Center that can provide this sort of information, unclassified sources include the reports of the International Crisis Group and the *SIPRI Yearbook*.

[8] By contrast, the process described in FM 3-24, *Counterinsurgency* (Headquarters, Department of the Army, December 2006) has the core issues and grievances flowing out of a basic data collection exercise. We think that the differences are primarily semantic in that we

lecting basic data on military, paramilitary, police, and insurgent numbers, weapons, and other capabilities, as well as collecting political, economic, social, and other data on national and sub-national groups and characteristics that may help to account for key fault lines, spatial patterning of violence, and other phenomena.[9]

In essence, this step aims to provide data that can assist the analyst in refining his understanding of major forces and fault lines that might explain factionalization, coalition formation, and other such phenomena. These data can speak to demographic, political, economic, social, ethnic, religious, sectarian, tribal, ideological, etc., fault lines; urban versus rural distinctions; and have and have-not distinctions. Data of interest include current national and sub-national snapshots, trend data, and forecasts related to civilian considerations.

Data of interest also include data on other key features of the operating environment, including its information domain (e.g., broadcast and print media infrastructure and audience analyses), along with opinion survey data—including, in some cases, respondent-level datasets—that may be available.[10]

The basic data that need to be collected for IW analysis are most often geospatially distributed (for example, the ethnic and sectarian composition and distribution of Iraq, and the location of improvised explosive devices or suicide attacks), so maintaining and displaying these data in a geospatial form can greatly facilitate analysis of IW environments. Figure 3.2, which was developed in a study of commanders'

---

break out core issues and grievances and use the data collection phase to connote efforts to collect quantitative data.

[9] A number of initialisms aim to capture this sort of information—for example, the C in METT-TC stands for civilian considerations, and ASCOPE stands for areas, structures, capabilities, organizations, people, and events.

[10] There are many sources for this sort of information, including, for example, the CIA's *World Factbook* (online at https://www.cia.gov/library/publications/the-world-factbook/); U.S. DoD Intelligence Production Program (DoDIPP) products; U.S. Army Civil Affairs country studies; Department of State country handbooks; Open Source Center media guides; the U.S. Bureau of the Census; DOS INR/Office of Research opinion research reports, and other polling; the United Nations, World Bank, and IMF; and the Statesman's Yearbook and Europa Yearbook.

Figure 3.2
Geospatially Oriented Aspects of the Information Domain of the
Operating Environment

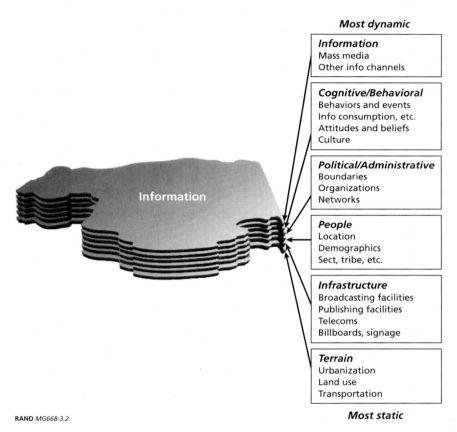

Most dynamic

Information
Mass media
Other info channels

Cognitive/Behavioral
Behaviors and events
Info consumption, etc.
Attitudes and beliefs
Culture

Political/Administrative
Boundaries
Organizations
Networks

People
Location
Demographics
Sect, tribe, etc.

Infrastructure
Broadcasting facilities
Publishing facilities
Telecoms
Billboards, signage

Terrain
Urbanization
Land use
Transportation

Most static

RAND *MG668-3.2*

information needs for influence operations, suggests that geospatially oriented data—in this case, data pertaining to the information domain of the operating environment—can fruitfully be displayed as a series of overlapping layers that facilitate the analysis of spatial correlations.[11]

As Figure 3.2 shows, features of the operating environment range from relatively static features of the terrain (e.g., urbanization, land use,

---

[11] This discussion draws from Larson et al., *Understanding Commanders' Information Needs for Influence Operations*, forthcoming.

and transportation networks) to more dynamic features. In a somewhat notional sense, these can include infrastructure, population distributions, and demographic or cultural characteristics; the prevailing cognitive traits of a given population in a specific region (e.g., attitudes and beliefs); population behaviors (e.g., attacks, protests, or useful tips on insurgents); and information, which is in a constant state of flux, with an ever-changing mix of new messages competing for attention at any given time. Organizing disparate sorts of data by location may, through visualization and spatial analysis, help to establish correlational patterns that otherwise might be masked, leading to fruitful insights about the dynamics of IW that might not otherwise occur to analysts.

Detailed Stakeholder Analyses

The second activity begins with the fifth step in the process, stakeholder characteristics. Here, the analyst builds on the earlier steps by adding detail about key characteristics of each stakeholder, be it an individual or a group.

At the highest level, these characteristics include the stakeholder's basic worldview, historical or cultural narrative, motivations, and views on key issues in contention; the importance or salience of the conflict or issue in dispute to the stakeholder; aims, objectives, preferred outcomes, and strategy; and morale, discipline, and internal cohesion or factionalization. They also include general and specific attitudes and beliefs related to the underlying conflict, as well as historical, cultural, religious, and linguistic characteristics, economic circumstances (e.g., income, unemployment rate), and other factors.

In this fifth step, the analyst also estimates each stakeholder's capabilities, both non-military and military. Non-military capabilities include the size of the stakeholder group (in terms of both raw numbers of members and estimates of the numbers of people it can mobilize or send into the streets) and its political, economic, and other non-military resources and capabilities.[12]

---

[12] These can include, for example, land and business ownership or entrepreneurial and organizational capabilities.

Another critical part of this step is making force assessments of each stakeholder's military, paramilitary, and other capabilities for undertaking violence. For the government, in addition to detailing conventional military organizations and their capabilities, force assessments must include various paramilitary, police, border, and other security forces. For the opposition, the assessments can include characterizations of militias, insurgent and terrorist groups, death squads, and other irregular organizations. In either case, detailed in this step are the estimated number of actual fighters associated with each stakeholder group or organization; basic organizational and order of battle (OOB) information; and estimates of readiness, discipline, effectiveness, penetration, corruption, and other factors that may affect performance. Also included are assessments of operational concepts used, including doctrine; tactics, techniques, and procedures (TTPs); leadership and organization; command, control, and communications (C3); and weapons system facilities (e.g., garrisons, weapons caches) related to organizations capable of employing violence. Finally—and especially for non-governmental organizations—it is important to understand the arms markets and networks that are the sources of weapons and systems.

The sixth step, stakeholder network and relationship/link assessment, involves a detailed analysis of formal organizational characteristics within and among groups, as well as informal links and networks, and the identification of leaders and influential individuals within the network. Formal organizational structures and relationships can be understood through the collection and analysis of organizational charts and tables of organization, and legal, administrative, and other materials can illuminate formal/legal authorities, control over resources, and other phenomena. Informal networks and relationships can involve people, domestic groups and institutions (e.g., banks, businesses), and external groups and institutions (e.g., states, transnational movements).

Thus, a second critical lens for unpacking the IW operating environment can be characterized in terms of overlapping or interlocking networks. This approach provides a view of a number of key features of the broader political society, including key leaders, their critical relationships, and their sources of authority, power, and influence. Networks can be used to characterize a host of formal organizations and hierarchies, whether they are political, military, bureaucratic, or administrative; economic or business-oriented; or tribal, religious, or sectarian. They also can be used to characterize informal networks, including personal and professional networks, networks characterizing patronage relationships or criminal enterprises, jihadist discourse, or influence. In addition, physical networks, such as telecommunications, command, control, communications, and computers (C4), and utilities, translate naturally into link and node data.[13]

Stakeholder leadership assessment, the seventh step, involves detailed leadership analyses where indicated. Such analyses tend to focus on key leaders. Assessments involve compiling and reviewing basic biographical information, as well as psychological profiles, assessments, and psychohistories; analyzing past decisionmaking for patterns; and carrying out other analyses that can illuminate individual-level motivations, aims, objectives, intentions, leadership preferences, pathologies, vulnerabilities, and decisionmaking styles, as well as connections to other individuals, groups, and places; favored communications channels; and other characteristics. Also important are the nature of bargains and social contracts between stakeholder leaders and followers (i.e., what leaders must provide to followers to retain their loyalty).

Dynamic Analyses

The final step in the IPB process is the integration of intelligence information to determine a threat's likely course of action (COA) and to understand the possible trajectory of the situation. We refer to these sorts of activities as *dynamic analyses.*

As witnessed in Iraq and Afghanistan, population-centric IW operations are conducted in troubled societies beset by intrastate political conflict, whether the origins of the conflict are to be found in economic, social, territorial, tribal, ethnic, sectarian, resource, or other grievances or differences, and whether the conflict takes the form of

---

[13] Additional detail on social network analysis can be found in Larson et al., *Understanding Commanders' Information Needs for Influence Operations*, forthcoming.

low-level terrorism or insurgency, or wide-scale civil war.[14] As many academic studies have shown, intrastate conflicts are difficult to stop. Rather than ending, most of them see temporary reductions in violence and death before restarting.[15] The usual explanation is that the combatants find it difficult to live together and cooperate in conditions of low trust, because they only recently were inflicting violence on each other. Put simply, IW environments can be quite dynamic, and it is critical to monitor a wide range of developments that can presage change and, where possible, to make forecasts regarding the possible future trajectory of these situations. That the different types of IW conflict and threats are often nested, linked, and simultaneous (e.g., insurgency coupled with terrorism) increases the challenges of dynamic analysis of IW.

As described earlier, the outcomes of IW environments are determined, first, by the capabilities and commitment of the government and its supporters (whether internal or external) relative to those of the government's challengers. But they also are determined by each side's willingness and ability to engage in political negotiations to build a coalition of supporters with enough capability to defeat or extract a compromise solution from the other side.

Analytic Techniques for Irregular Warfare Analysis

A number of analytic techniques may be helpful to intelligence analysts seeking to understand or anticipate the path of an IW situation.

Agent-based rational choice or expected utility models. A family of models—agent-based rational choice models—has been developed to provide computationally based forecasts of complex, multi-actor, real-world political issues such as IW situations. These models incorporate insights from spatial politics, social choice theory, game theory, and expected utility theory in a form that enables policy-relevant fore-

---

[14] To be sure, and as will be discussed, other states can be important stakeholders and seek to influence internal conflicts; the distinction is relative to interstate, state-versus-state conflicts.

[15] See Harbom, Hogbladh, and Wallensteen, "Armed Conflict and Peace Agreements, *Journal of Peace Research*, 2006, for a detailed discussion of which armed conflicts abated in 2005 and which restarted.

casts based on fairly modest data inputs. Even more important, some forms of these models have an impressive record of predicting the outcome of a wide range of political phenomena—including conflict—with an order of 90 percent accuracy.[16]

Perhaps the most prominent feature of these models from the standpoint of assessing IW environments is that they enable dynamic forecasts based on a relatively small subset of the factors identified in our analytic framework:

- the existence of many different stakeholder groups that may seek to influence the outcome of the contest between the government and its challengers
- the possibility that different stakeholder groups may have different grievances or objectives, or take different positions on various issues related to the contest between the government and its challengers
- differing relative political, economic, military, organizational, and other capabilities of stakeholder groups
- differences in the perceived importance of and level of commitment to the dispute for each stakeholder group, with some potentially viewing the stakes as existential while others remain disengaged or indifferent.[17]

---

[16] See, for example, Bruce Bueno de Mesquita, "The Methodical Study of Politics," paper, October 30, 2002; and James Lee Ray and Bruce Russett, "The Future as Arbiter of Theoretical Controversies: Predictions, Explanations and the End of the Cold War," *British Journal of Political Science*, Vol. 26, 1996, pp. 441–470. A detailed discussion of these models, and a more complete review of claims about their predictive accuracy, can be found in Larson et al., *Understanding Commanders' Information Needs for Influence Operations*, forthcoming.

[17] These models rest on several other assumptions, including: (1) the assumption that the overall outcome of a conflict typically will be the position of the median stakeholder, when each stakeholder is weighted by his relative effective capabilities; (2) the assumption that the path to an outcome can be characterized by any mix of bargaining and/or conflict between the groups, and that, as a result, stakeholder groups' positions, level of commitment, or capabilities can change; and (3) the assumption that even after an outcome is determined, some stakeholders may continue to oppose that outcome.

In this view, the ultimate question for the analyst conducting a dynamic assessment of an IW environment is the nature of the political equilibrium outcome that is forecast and whether that equilibrium outcome meets U.S. policy objectives. In some, perhaps most, cases, the predicted equilibrium may be well short of what the United States is hoping to accomplish. In such cases, sensitivity analyses can illuminate the combination of factors that might be required to achieve U.S. objectives—or, indeed, whether the United States can plausibly achieve its objectives at all—and where the greatest leverage for influencing the equilibrium outcome lies.[18] Periodic updates to such dynamic assessments also are likely to be required to ensure that the assumptions and inputs used in the baseline assessment remain current.

We also note that although we discuss these models in the context of the third set of analytic activities, they can, in fact, be used early in the analytic process to help illuminate which stakeholder groups are the most influential or otherwise important in determining the outcome of an IW situation and thus should be accorded higher levels of analytic attention.

Analytic tools for IW analysis identified in doctrine. Available Army doctrine identifies a number of analytic techniques and tools suitable for IW analysis, some of which we have already discussed in the context of our analytic framework.[19] These include

- link analysis/social network analysis, which can be used to understand critical links between individuals, institutions, and other components

---

[18] In other cases, the forecast outcome of the conflict may be even more favorable from a U.S. perspective than the U.S. limited objectives are. In these cases, policymakers and strategies would be faced with the happy choice of pursuing these more favorable objectives, or tailoring U.S. strategy or reducing the level of effort to what is needed to achieve current objectives.

[19] Appendix B of FM 3-24 (Headquarters, Department of the Army, *Counterinsurgency*, 2006) and Appendix B of FM 3-06 (Headquarters, Department of the Army, *Urban Operations*, Washington D.C., October 2006) each provide a useful summary and description of these analytical techniques.

- pattern analysis, which can illuminate temporal or spatial patterning of data and provide a basis for insights into underlying correlational or causal mechanisms that can be used to evaluate a threat and to assess threat COAs
- cultural comparison matrixes, which can help to highlight similarities, differences, and potential points of congruity or friction between groups
- historical timelines, which list significant dates and relevant information and analysis that can be used to underwrite a larger historical narrative about the sources of grievances, onset of violence, and other phenomena, as well as provide insights into how key population segments may react to certain events or circumstances
- perception assessment matrixes, which can be used to characterize the cultural lenses different groups use in viewing the same events
- spatial analysis/map overlays, which can be used to assess spatial relationships or correlations between disparate geographically distributed characteristics
- psychological profiles, which can assist in understanding how key groups, leaders, and decisionmakers perceive their world.[20]

Additionally, trend analyses—a form of pattern analysis—may be a particularly fruitful approach for IW analysts. Whether focused on time series data describing significant activities (SIGACTs), changing media content or population attitudes, or exploring correlations between disparate variables, trend analyses can help further illustrate dynamic processes.

Other diagnostic models. In addition to various worthwhile scholarly efforts that have systematically addressed dynamic aspects of intrastate violence, there are several other policy-relevant diagnostic tools

---

[20] A detailed list of doctrinal publications that address the analysis of IW can be found in Appendix B.

that either share some features of our analytic framework or accent somewhat different phenomena that may be useful to IW analysts.[21]

Anticipating intrastate conflict. Because early diplomatic, military, or other policy action can in some cases reduce the prospects of full-blown conflict emerging, intelligence analysts sometimes require tools for anticipating intrastate conflict.[22] An earlier RAND study developed a process model for anticipating intrastate conflict (focusing on ethnic and sectarian conflict, though readily applicable to other types of intrastate conflict) and a handbook of questions and guidelines for analysts that revolved around five analytic steps:

- *Identify structures of closure.* In this step, the analyst identifies structural factors that close off political, economic, or social opportunities for stakeholder groups and may thereby lead to strife.
- *Map closure onto identifiable affinities.* In this step, the analyst identifies which stakeholder groups—whether based on kinship, race, language, religion, region, culture, or some other factor— are facing which types of closure.
- *Identify catalysts of mobilization.* In this step, the analyst identifies factors that may mobilize excluded stakeholder groups—e.g., a change in the balance of power, "tipping events," the emergence of

---

[21] Among the more noteworthy scholarly efforts are Charles Tilly's *From Mobilization to Revolution* (New York: McGraw-Hill, 1978) and *Politics of Collective Violence* (Cambridge: Cambridge University Press, 2003); James DeNardo's *Power in Numbers: The Political Strategy of Protest and Rebellion* (Princeton: Princeton University Press, 1985); and Sidney Tarrow's *Power in Movement: Social Movements and Contentious Politics* (Cambridge: Cambridge University Press, 1998).

[22] As Thomas Szayna and Ashley Tellis put it: "Dealing with the consequences of communitarian conflicts is not an optimal way to address the problem of ethnic strife. A better understanding and anticipation of such conflicts, which consequently improves the prospects for preemptive remedial action short of using forces, is a much better alternative. In short, preventing strife is almost always a more efficient strategy than dealing with the consequences of strife. . . . [And] for reasons of preventing long-term strife that may escalate to major regional problems, prevention is the preferred course of action." Thomas S. Szayna and Ashley J. Tellis, "Introduction," in Thomas S. Szayna, ed., *Identifying Potential Ethnic Conflict: Application of a Process Model*, MR-1188-A, Santa Monica, Calif.: RAND Corporation, 2000, pp. 3, 5.

policy entrepreneurs who seek to exploit dissatisfaction, increased resources and improved organization, and external assistance.

- *Assess state capability.* In this step, the analyst assesses the state's political capacity to accommodate aggrieved stakeholder groups, its fiscal capacity to compensate them, and its coercive capacity to suppress them.
- *Forecast likelihood of violence.* In this step, the analyst estimates, based on an analysis of the government and its opponents, the likelihood of political conflict using game theoretical reasoning.[23]

As just described, this model is a diagnostic tool for considering the motivations various stakeholder groups might have for challenging a government, and for assessing the government's relative capabilities for avoiding escalation by accommodating or suppressing its challengers before they can mobilize and undertake mass violence. While not predictive of intrastate violence, this model can help assess whether the conditions for such violence are present or not, improve the analyst's understanding of the drivers of conflict, and point out data needs and limitations.

Trigger and risk factors for religious groups choosing violence. Work done by RAND colleague Greg Treverton on the analysis of religious groups identified five potential triggers and risk factors for violence that had some interesting parallels to our conception of dynamic IW analysis:

- *Belief in victory.* Belief that the use of force can achieve the desired political end encourages violence.
- *Fear of annihilation.* Existential threats can cause and sustain violence.
- *Inability or unwillingness to participate in politics.* Being blocked from or uninterested in "normal" politics leaves force as the other option for pursuing goals.

---

[23] For a detailed exposition of this model, see Szayna, *Identifying Potential Ethnic Conflict: Application of a Process Model*, 2000, pp. 30–73 and appendix titled "Questions and Guidelines for the Analyst," pp. 291–328.

- *Young and inexperienced leadership.* Youthful leadership is sometimes risk taking and inexperienced, and in crisis situations may aggressively lead a group into violence.
- *Political and economic crisis.* Economic collapse combined with political crisis enhances the ability of religious groups to wage war by increasing their ideological and material appeal.[24]

Counterterrorism Operations

Policy and strategy guidance for the war on terrorism is provided in the *Quadrennial Defense Review Report*, the *National Strategy for Combating Terrorism*, and the *National Military Strategic Plan for the War on Terrorism* (NMSP-WOT), the principal policy and strategy documents related to U.S. government counterterrorism activities.[25] We also note that NMSP-WOT 2/06 distinguishes between counterterrorism operations conducted under the war on terrorism theater campaign plans of the geographic combatant commands in theaters such as Iraq and Afghanistan, and those conducted under SOCOM's global campaign plan for the war on terrorism.[26]

Our review of existing doctrine suggests that it tends to treat terrorism and insurgency as largely identical phenomena and does not differentiate between the intelligence requirements for counterinsurgency and counterterrorism operations. Thus, although the intelligence analytic requirements of a global jihadist insurgency are somewhat less distinct than those of typical insurgencies, counterterrorism operations do appear to share many of the analytic requirements of the population-

---

[24] Private communication from Gregory F. Treverton, March 30, 2007.

[25] DoD, *Quadrennial Defense Review Report*, 2006; White House, *National Strategy for Combating Terrorism*, Washington, D.C., September 2006; and Chairman of the Joint Chiefs of Staff, *National Military Strategic Plan for the War on Terrorism*, Washington, D.C., February 1, 2006 (from here on referred to as NMSP-WOT 2/06). For a review of recent policy and strategy documents that address IW, see Appendix A.

[26] NMSP-WOT, 2/06, p. 9. Thus, NGIC's analytic requirements theoretically could originate either from SOCOM or from the geographic combatant command or a subordinate command (e.g., Multi-National Corps–Iraq).

centric IW environments discussed earlier. For example, terrorism—the terrorizing of a civilian population—is an extreme form of coercing and influencing a government or population, the success of which is susceptible to analysis using the framework for population-centric IW situations. Put another way, like insurgents, terrorists compete for the support or compliance of the larger population:

> Extremists use terrorism—the purposeful targeting of ordinary people—to produce fear to coerce or intimidate governments or societies in the pursuit of political, religious, or ideological goals. Extremists use terrorism to impede and undermine political progress, economic prosperity, the security and stability of the international state system, and the future of civil society.[27]

In addition, terrorists' actions play to an audience of their own supporters, demonstrating the terrorists' ability to effectively conduct operations. In this way, they enhance morale and support.

Terrorist networks also share many of the conceptual features of other adversary networks, including insurgent networks, that are already the subject of detailed intelligence analysis for targeting and other purposes:

> All enemy networks rely on certain key functions, processes, and resources to be able to operate and survive. These three elements are an important basis for counter-network strategies and can be defined as follows:
> — *Function* (Critical Capability): A specific occupation, role, or purpose.
> — *Process*: A series of actions or operations (i.e., the interaction of resources) over time that bring about an end or results (i.e., a function).
> — *Resource* (Critical Requirement): A person, organization, place, or thing (physical and non-physical) and its attributes. In network vernacular, a resource may also be referred to as a "node"

---

[27] NMSP-WOT 2/06, p. 4.

and the interaction or relationship between nodes described as "linkage."[28]

NMSP-WOT 2/06 helpfully provides a categorization of the elements of a terrorist network that is meant to serve "as a common lexicon for orienting and coordinating efforts against enemy networks," and as a framework for the analysis of a network's critical elements for operation and survival. Although this thinking is not yet enshrined in doctrine, it may nevertheless represent a starting point for counterterrorism analysis.

According to NMSP-WOT 2/06, terrorist and other adversary networks comprise nine basic components:

- leadership
- safe havens
- finance
- communications
- movement
- intelligence
- weapons
- personnel
- ideology.[29]

The elements in this list are quite similar to many of those identified in our presentation of the IW analytic framework as it applies to counterinsurgency and other population-centric IW situations, which were discussed in some detail in the earlier description of our IW analytic framework.

It also is worth mentioning in this connection David Kilcullen's work, which treats counterinsurgency as a "complex system" and the larger war on terrorism as a "global counterinsurgency."[30] Moreover,

---

[28] NMSP-WOT 2/06, pp. 4–5.

[29] NMSP-WOT, 2/06, pp. 14–19.

[30] See David Kilcullen, "Countering Global Insurgency," *Small Wars Journal*, 2004 (online as of September 2008 at http://www.smallwarsjournal.com/documents/kilcullen.pdf). Kilcullen, an Australian, reportedly helped shape the QDR's treatment of the war on terrorism

there are no apparent inconsistencies between Kilcullen's approach, which focuses on key nodes, links, boundaries, interactions, subsystems, inputs, and outputs, and our analytic framework. Although Kilcullen's application of complex systems theory appears still to be embryonic, he has written a number of interesting papers dealing with counterinsurgency and the war on terrorism that may prove useful for IW analysts and may suggest research directions deserving of further exploration.[31]

That said, there are some features of counterterrorism intelligence requirements that differ from population-centric IW and bear discussion. We next describe features associated with two different categories of counterterrorism operations—tactical counterterrorism operations, and operations against transnational terrorist networks—that might lead to some slight differences in intelligence analytic requirements.

Tactical Counterterrorism Operations

From a strict doctrinal perspective, counterterrorism is a SOF mission, typically involving direct action by SOF. Mission doctrine and intelligence requirements are the responsibility of the special operations community. Most of this doctrine is not available to the public.

Nevertheless, operations against terrorist cells can take place in the context of conventional or IW campaigns, and the IW JOC 9/07 explicitly discusses the likelihood of operations against terrorists conducted as part of larger, counterinsurgency campaigns.[32] For exam-

---

when he was Chief Strategist in the Office of the Coordinator for Counterterrorism at the U.S. State Department. See, also, George Packer, "Knowing the Enemy," *The New Yorker*, December 18, 2006.

[31] On the subject of classic counterinsurgency, see Kilcullen's "Counterinsurgency *Redux*" (*Small Wars Journal*, online edition, undated) and "Twenty-Eight Articles: Fundamentals of Company-Level Counterinsurgency" (*Iosphere*, Summer 2006, pp. 29–35). On the subject of the global war on terrorism as a "global insurgency," see Kilcullen's "Countering Global Insurgency" (*The Journal of Strategic Studies*, Vol. 28, No. 4, August 2005, pp. 597–617). Perhaps the most detailed treatment of Kilcullen's application of systems thinking to insurgency is in Appendix C, "Case Study—Systems Assessment of Insurgency in Iraq," of an earlier version (Version 2.2) of "Countering Global Insurgency," dated November 30, 2004.

[32] For example, the Multi-National Corps–Iraq conducts operations against the local Al Qaeda affiliate, the Al Qaeda Jihad Organization in the Land of Two Rivers.

ple, cordon and search operations can flush out individual terrorists or small cells from their safe havens; conventional ground forces can conduct targeted raids against terrorist targets; or longer-range strike capabilities, such as aircraft, cruise missiles, and unmanned combat aerial vehicles, can be employed. In most of these circumstances, standard doctrine for conventional tactical operations applies, as does extant doctrine on intelligence analysis for these operations. Nevertheless, the success of these operations hinges to a great degree on the timeliness, precision, and responsiveness of intelligence.

In either case, at the operational level, as with population-centric IW environments such as counterinsurgency, such factors as safe houses, enclaves of popular support, arms smuggling networks, networks for recruitment and training, weapons caches, and other phenomena are of great interest to the intelligence analyst.

Operations Against Transnational Terrorist Networks

By comparison, and largely for reasons of classification, the intelligence analytic requirements of the United States' broader strategy for the greater war on terrorism are less well developed in the open literature.[33] The unclassified NMSP-WOT does, however, list a number of annexes that suggest a number of discrete counterterrorism activities, each of which would be presumed to have associated with it a set of intelligence and analytic requirements. These include

- *Annex B, Intelligence:* Describes the threat, concept of intelligence operations, and intelligence activities.
- *Annex C, Operations:* Provides tasks and coordinating instructions for implementing the base plan, and assigns DoD-wide responsibilities. It establishes Commander of the United States Special

---

[33] The classified version of the NMSP-WOT includes an annex describing the threat, concept of intelligence operations, as well as intelligence activities (NMSP-WOT 2/06, p. 28). The classified SOCOM Global Campaign Plan for the War on Terrorism document would be expected to also discuss intelligence requirements of the greater war on terrorism (NMSP-WOT, 2/06, p. 9). A next step for further refinement of intelligence requirements for the greater war on terrorism would be to review these documents.

Operations Command (CDRUSSOCOM) as the supported commander in the greater war on terrorism.

- *Annex F, Public Affairs:* Coupled with the strategic communications guidance in Annex H, offers guidance for implementing public affairs campaigns in support of the greater war on terrorism.
- *Annex H, Strategic Communication:* Coupled with the public affairs guidance in Annex F, offers guidance for implementing strategic communications campaigns in support of the greater war on terrorism.
- *Annex L, Homeland Defense, Homeland Security, and Civil Support:* Describes the military's role in each mission area as it relates to the greater war on terrorism.
- *Annex T, WMD/E:* Describes weapons of mass destruction/effects (WMD/E) terrorism and the principles for combating WMD/E.[34]

Thus, it is relatively easy to imagine IW intelligence analysts being asked to conduct operationally relevant analyses in support of ground commanders, either to provide analytic support to public affairs or strategic communications activities directed at countering ideological support to terrorism, or to provide assessments related to WMD or other dangerous substances or technologies.

Nevertheless, SOCOM and the geographic combatant commands are likely to be the best sources for identifying their own unmet intelligence needs.

Comparison to the Standard IPB Process

Doctrinally, the purpose of the IPB process is to systematically and continuously analyze the threat and environment in a specific geographic area in order to support military decisionmaking, enabling the

---

[34] See NMSP-WOT, 2/06, pp. 28–33.

commander to selectively apply his combat power at critical points in time and space. This process consists of four steps:

- *Defining the operational environment:* In this step, the analyst seeks to identify for further analysis and intelligence collection the characteristics that will be key in influencing friendly and threat operations.
- *Describing the operational environment:* In this step, the analyst evaluates the effects of that environment on friendly and threat forces. It is in this step that limitations and advantages that the operational environment provides for the potential operations of friendly and threat forces are identified.
- *Evaluating the threat:* In this step, the analyst assesses how the threat normally operates and organizes when unconstrained by the operational environment. This step is also used to identify high-value targets.
- *Determining threat COA:* In this step, the analyst integrates information about what the threat would prefer to do with the effects of the operational environment in order to assess the threat's likely future COA.

We can assess the compatibility of our analytic framework with the IPB process by examining the coincidence of activities in Table 3.1, which compares elements of our analytic framework (the rows) with the standard doctrinal IPB process (the columns).[35]

When one looks down the columns of the table, it should be clear that our analytic framework involves activities that are conducted under each step of the four-step IPB process. For example, three of the steps in our framework's preliminary assessment and basic data collection phase are congruent with the first step of the standard doctrinal IPB process, and three are congruent with the IPB process's second step. The reason for this congruence is that existing Army doctrine

---

[35] The IPB process is described in detail in Headquarters, Department of the Army (HQDA), *Intelligence Preparation of the Battlefield*, FM 34-130, July 1994. The application of the IPB process during a COIN operation can be found in FM 3-24 (Headquarters, Department of the Army, *Counterinsurgency*, pp. 3-2 to 3-24).

Table 3.1
Crosswalk with Standard IPB Process

Analytic Framework Activities	Standard IPB Process Steps			
	Define Operational Environment	Describe Operational Environment	Evaluate Threat	Determine Threat COA
1. Initial assessment and data gathering				
a. Preliminary assessment	X	X		
b. Issue/grievance identification		X		
c. Stakeholder identification	X			
d. Basic data collection	X	X		
2. Detailed stakeholder analyses				
a. Stakeholder characteristics		X	X	
b. Key leader analysis			X	
c. Stakeholder link analysis				
3. Dynamic assessment			X	X

fully supports the gathering and analysis of extensive information on the civilian and societal characteristics of the area of operations.

If one looks down the diagonal from the top left of Table 3.1 to the bottom right, it becomes clear that the basic ordering of analytic activities in our framework correlates fairly well with that of the standard IPB process.

In summary, a comparison of our framework with the standard IPB process suggests a high level of compatibility between the two and the possibility that our framework—or discrete elements of it—might be incorporated into existing intelligence analytic processes and educational and training curricula. In consequence, our analytic framework might best be viewed not as an alternative or competitor to IPB, but as providing an efficient analytic protocol for IW IPB analysis, one that is suitable for operational- and strategic-level intelligence analysis and that complements the IPB process's tactical-operational focus.

Chapter Conclusions

Based on our analysis of available doctrinal and other materials, we have presented an analytic framework, or procedure, that can be used by NGIC analysts to analyze IW environments—whether they entail "population-centric IW" situations potentially involving large-scale mobilization of the populace (such as a counterinsurgency), or address the "global jihadist insurgency" represented by the Al Qaeda ideological movement and network.

CHAPTER FOUR

Conclusions

The aim of this study was to develop an analytic framework for assessing IW situations that could subsequently be used as the basis of an educational and training curriculum for intelligence analysts charged with assessing IW situations.

The framework we developed takes the form of an analytic procedure, or protocol, consisting of three main activities—initial assessment and data gathering, detailed stakeholder analyses, and dynamic analyses—that involve eight discrete analytic steps. The central idea is that this is an analytic procedure by which an analyst, beginning with a generic and broad understanding of a conflict and its environment, can then engage in successively more-focused and more-detailed analyses of selective topics to develop an understanding of the conflict and to uncover the key drivers behind such phenomena as orientation toward the principal protagonists in the conflict, mobilization and recruitment, and choice of political bargaining or violence. Put another way, the framework can be used to efficiently decompose and understand the features of IW situations—whether of the population-centric or the counterterrorism variety—by illuminating areas in which additional detailed analysis could matter and areas where it probably will not matter.

As described in Chapter Three, our framework—and its constituent analytic activities and tools—is compatible with the military IPB process and its supporting analytic techniques. The framework also shares some characteristics of other policy-relevant models that have been developed as diagnostic tools for different purposes—e.g., anticipating ethnic conflict or assessing the prospects that religious groups will choose to resort to violence. Nonetheless, just as no two IW situations are identical, we envision that each new case will require some adaptation or refashioning of this framework.

We think that our analytic framework's greatest, albeit indirect, value may be to forces in the field in that it can provide the intelligence analysts who are supporting those forces with a construct for analyzing critically important but analytically vexing strategic- and operational-level features of IW situations. The framework also may be of value, in this case more direct and limited, in providing IW intelligence analysts in the broader intelligence community with a common frame of reference and language for community efforts to support policy, strategy, and operational design for IW situations.

A potential next step would be to test our analytic framework by applying it to an ongoing IW situation, such as Iraq or Afghanistan, or to a situation in which U.S. forces are not directly engaged, such as Lebanon. An application of this sort would provide a wealth of practical examples that could be used in the development of educational and training curricula and could help identify areas of weakness in the framework that require further refinement.

A Review of Defense Policy, Strategy, and Irregular Warfare

The growing importance of IW to the defense community, which is largely a result of the U.S. strategy to deal with global jihadists, and the range of specific challenges the United States has encountered in the Afghan and Iraqi insurgencies, have led to a high level of policy- and strategy-level attention to the requirements of IW. We briefly review recent policy- and strategy-level developments related to DoD thinking on IW.

The *National Defense Strategy of the United States of America* and *National Military Strategy of the United States of America* of March 2005 divided threats into four major categories: traditional, irregular, disruptive, and catastrophic.[1] In the view of these documents, the principal irregular challenge was "defeating terrorist extremism," but counterinsurgencies, such as those faced in Afghanistan and Iraq, were also included.

The *National Defense Strategy* also identified terrorism and insurgency as being among the irregular challenges the United States faces, the dangers of which had been intensified by two factors: the rise of extremist ideologies and the absence of effective governance. It described irregular threats as challenges coming "from those employing 'unconventional' methods to counter the *traditional* advantages of

---

[1] DoD's *National Defense Strategy of the United States of America* (Washington, D.C., March 2005, pp. 2–4) and *National Military Strategy of the United States of America: A Strategy for Today; A Vision for Tomorrow* (Washington, D.C., March 2005, p. 4).

stronger opponents" [emphasis in original],[2] and identified "improving proficiency for irregular warfare" as one of eight JCAs that would provide a focus for defense transformation efforts.[3] The *National Military Strategy* reprised many of the same points.[4]

The February 2006 QDR also identified IW as an emerging challenge:

> The enemies in this war are not traditional conventional military forces but rather dispersed, global terrorist networks that exploit Islam to advance radical political aims. These enemies have the avowed aim of acquiring and using nuclear and biological weapons to murder hundreds of thousands of Americans and others around the world. They use terror, propaganda and indiscriminate violence in an attempt to subjugate the Muslim world under a radical theocratic tyranny while seeking to perpetuate conflict with the United States and its allies and partners. This war requires the U.S. military to adopt unconventional and indirect approaches.[5]

> In the post-September 11 world, irregular warfare has emerged as the dominant form of warfare confronting the United States, its allies and its partners; accordingly, guidance must account for distributed, long-duration operations, including unconventional warfare, foreign internal defense, counterterrorism, counterinsurgency, and stabilization and reconstruction operations.[6]

The QDR operationalized strategy in terms of four priority areas for QDR examination: defeating terrorist networks, defending the homeland in depth, shaping the choices of countries at strategic crossroads,

---

[2] DoD, *National Defense Strategy*, 2005, p. 2. By September 2006, a third factor had been added: the potential for extremists to acquire WMD.

[3] DoD, *National Defense Strategy*, 2005, p. 3.

[4] DoD, *National Military Strategy*, 2005, pp. 4, 23.

[5] DoD, *Quadrennial Defense Review Report*, 2006, p. 1.

[6] DoD, *Quadrennial Defense Review Report*, 2006, p. 36. The QDR mentions "irregular" 41 times and "irregular warfare" 19 times.

and preventing hostile states and non-state actors from acquiring or using WMD (see Figure A.1).[7]

The QDR also refined DoD's force planning construct, dividing DoD military activities into three types of campaigns: conventional campaigns, war on terrorism/irregular (or asymmetric) warfare, and homeland defense (see Figure A.2).[8]

As described in this construct, senior DoD policymakers envisaged IW and the war on terrorism to include a broad range of activities, ranging from "active partnership and tailored shaping" and training and equipping foreign forces, to information operations, interdiction,

Figure A.1
2006 Quadrennial Defense Review's Priorities

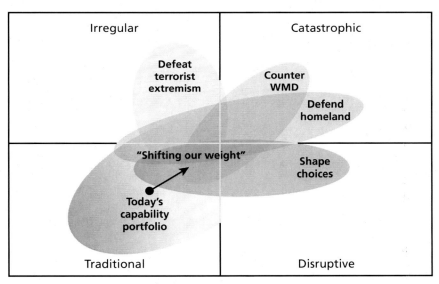

SOURCE: Joint Staff J-7 JETCD, "CJCS JCA Progress Report; SECDEF Action Memo Tasks; 1-Year Update, 24 August 2006," briefing, Washington, D.C., August 2006, slide 3.
RAND *MG668-A.1*

---

[7] DoD, *Quadrennial Defense Review Report*, 2006, pp. 19–35.

[8] DoD, *Quadrennial Defense Review Report*, 2006, pp. 35–39.

Figure A.2
2006 Quadrennial Defense Review's View of Threats

RAND MG668-A.2

foreign internal defense, stability operations, counterinsurgency, and even WMD elimination.

It is clear from the QDR's discussion that IW campaigns can take innumerable forms, but that these forms are, in a practical sense, bounded by two distinct, if stylized, ideal types that are of the greatest policy interest.

The first stylized ideal type is IW against "dispersed, global terrorist networks that exploit Islam to advance radical political aims," and focuses on the Al Qaeda organization umbrella of ideologically connected, cellular-structured groups. This form of IW, which targets specific individuals or small cells widely dispersed across the globe, requires an exquisite level of precision and timeliness in intelligence, targeting, and striking capability. It is highly tactical and technical in nature and generally does not rely on general-purpose forces; rather, the principal application of military power consists of direct action by small numbers of SOF and, presumably, precision strike.[9]

The second stylized ideal type of IW engenders more typical counterinsurgency operations, such as those being conducted in Iraq and Afghanistan. These can involve a much wider range of activities conducted by, or synchronized with, potentially large numbers of U.S. conventional forces. To be sure, these activities can include combat operations, but actual success more often depends on political efforts to reach a stable political equilibrium, underwritten by improvements to personal security for the population, as well as restoration of essential services, and economic development and good governance.[10] In this form of IW, the focus is less on military than on political, psychological, informational, and related efforts—less on defeating enemy forces than on persuading those who can be persuaded to support the government supported by the United States.

Finally, the QDR called for a number of efforts to improve the U.S. military's IW capabilities. Importantly, these included a rebalancing of general-purpose forces by shifting the focus from conventional

---

[9] Precision strike includes the MQ-9 Reaper (formerly the RQ-1 Predator B) unmanned combat aerial vehicle, ground attack aircraft, long-range bombers, and cruise missiles. Action by civilian law enforcement or paramilitary capabilities also could be used to capture or incapacitate terrorists. We also note requirements for countering ideological support for terrorism (CIST)—i.e., the larger, "war of ideas" that aims to reduce support among Muslims for extremist positions. We believe that while this may be critically important military activity in a ground commander's area of operations, the global campaign generally is less of a military responsibility than a civilian one.

[10] According to Joint Chiefs of Staff, "Irregular Warfare (IW) Execution Roadmap," undated: "Tactical and Operational competence in conventional warfighting does not necessarily guarantee tactical, operational, or strategic success in operations and activities associated with IW."

warfare to IW while boosting the ranks of SOF.[11] It also included working with foreign governments to enhance the capacity of their militaries and security forces for dealing with threats within their borders or in their regions, and forging closer cooperation with other parts of the federal government involved in the war on terrorism.

Another document, NMSP-WOT 2/06, identified six objectives for the global war on terrorism: (1) deny terrorists the resources they need to operate and survive; (2) enable partner nations to counter terrorist threats; (3) deny WMD technology to U.S. enemies and increase capacity for consequence management; (4) defeat terrorist organizations and networks; (5) counter state and non-state support for terrorism in coordination with other U.S. government agencies and partner nations; (6) counter ideological support for terrorism.[12]

In September 2006, the White House released an updated version of its *National Strategy for Combating Terrorism* that identified the following elements:

- advance effective democracies as the long-term antidote to the ideology of terrorism
- prevent attacks by terrorist networks
- deny WMD to rogue states and terrorist allies who seek to use them
- deny terrorists the support and sanctuary of rogue states
- deny terrorists control of any nation they would use as a base and launching pad for terror
- lay the foundations and build the institutions and structures needed to carry the fight forward against terror and help ensure ultimate success.[13]

---

[11] DoD, *Quadrennial Defense Review Report*, 2006; the enhancement of general-purpose forces' IW capabilities is described on pp. 38, 44; SOF expansion is described on pp. 5 and 43–45.

[12] A classified version of this document was released at the same time that superseded the NMSP-WOT of October 19, 2002 (NMSP-WOT 2/06, p. 28).

[13] White House, *National Strategy for Combating Terrorism*, 2006, p. 1. The earlier version was released in February 2003.

Subsequent to the QDR's release, execution roadmaps were developed for each of the eight JCAs identified in the QDR to convert the broad policy objectives established in the QDR to actionable tasks and program objectives memorandum (POM) guidance for the Fiscal Years 2008–2013 Future Years Defense Program. The IW execution roadmap, which was completed in the spring of 2006, assessed opportunities to implement DoD Directive (DODD) 3000.05, DoD's December directive on SSTRO, as well as ways to improve the military's ability to counter a long-term guerilla war.[14]

According to testimony, the classified version of the execution roadmap identified about 30 tasks for improving DoD's proficiency for IW,[15] along with the following five major initiatives for implementation in the 2008–2013 defense program: (1) change the way DoD manages the people necessary to support IW; (2) rebalance general-purpose forces to better support IW; (3) increase SOF capabilities and capacity to support IW; (4) increase capacity to conduct counter-network operations; (5) redesign joint and service education and training programs to conduct IW.[16]

In April 2006, Deputy Secretary of Defense Gordon England directed the Marine Corps and SOCOM to develop an "Irregular Warfare Joint Operating Concept," or JOC, the purpose of which

---

[14] See Mario Mancuso, "Irregular Warfare Roadmap," *Special Operations Technology*, online edition, January 2007; and Under Secretary of Defense for Policy, "Military Support for Stability, Security, Transition, and Reconstruction (SSTR) Operations," DODD 3000.05, November 28, 2005. According to press reporting, the IW execution roadmap was led by Principal Deputy Under Secretary of Defense for Policy Ryan Henry, and Joint Staff Director of Operations and Marine Corps Lt. Gen. James Conway (Jason Sherman, "DoD Plans for Life After the QDR," InsideDefense.com NewsStand, January 12, 2006). The classified execution roadmap reportedly was signed by Deputy Defense Secretary Gordon England on April 28, 2006 (Jason Sherman, "New Blueprint for Irregular Warfare," InsideDefense.com NewsStand, May 16, 2006).

[15] Two other companion roadmaps, the Building Partnership Capacity Roadmap and the Strategic Communications Roadmap, also were commissioned.

[16] Statement by Mr. Mario Mancuso, Deputy Assistant Secretary of Defense for Special Operations and Combating Terrorism, Before the 109th Congress, Committee on Armed Services, Subcommittee on Terrorism, Unconventional Threats and Capabilities, United States House of Representatives, September 27, 2006.

was to "broadly describe how joint force commanders will conduct protracted IW to accomplish national objectives on a regional level or global scale."[17]

The IW JOC 9/07 argued that IW was likely to become an increasing challenge for the U.S. Government:

> Our adversaries will pursue IW strategies, employing a hybrid of irregular, disruptive, traditional, and catastrophic capabilities to undermine and erode the influence and will of the United States and our strategic partners. Meeting these challenges and combating this approach will require the concerted efforts of all available instruments of U.S. national power. . . . This concept describes IW as a form of warfare and addresses the implications of IW becoming the dominant form of warfare, not only by our adversaries but also by the United States and its partners.[18]

IW has a dual nature, including both offensive and defensive aspects and approaches:

> Many factors may preclude or restrain a joint force from conducting conventional military campaigns. These situations will require or favor an irregular military approach, using indirect and often non-traditional methods to achieve U.S. strategic objectives. IW will become an increasingly attractive strategic option and perhaps a preferred means for the U.S. to influence, deter, or defeat hostile states, occupying powers, and non-state adversaries. At the same time, the defensive use of IW will help keep in check those who wish to do us or our friends or allies harm.[19]

Finally, the IW JOC aimed to provide the framework for developing IW-related capabilities via POM submissions:

---

[17] Statement of Vice Admiral Eric T. Olson, Deputy Commander, United States Special Operations Command, Before the House Armed Services Committee, Subcommittee on Terrorism, Unconventional Threats and Capabilities, On Irregular Warfare, September 27, 2006.

[18] IW JOC 9/07, p. 1.

[19] Olson, Statement of Vice Admiral Eric T. Olson, September 27, 2006.

The IW concept will apply across the full range of military operations and will result in a more balanced DOD approach to conflict. The concept will also provide guidance for force development that could result in changes to doctrine, organization, training, materiel, leader development and education.[20]

The JOC will guide the development and integration of Department of Defense (DOD) military concepts and capabilities for waging protracted IW on a global or regional scale against hostile states and armed groups. The JOC will provide a basis for further IW discussion, debate, and experimentation intended to influence subsequent IW concept and capability development. It will also influence joint and Service combat development processes by helping the joint force gain a better appreciation for IW challenges that will result in doctrine, organization, training, materiel, leadership and education, personnel, and facilities (DOTM-LPF) changes.[21]

According to testimony given during a September 2006 House Armed Services Committee hearing on the IW roadmap, the essence of IW is non-military in nature:

While the precise definition continues to be refined, there is broad agreement within the Department on the nature and scope of Irregular Warfare. What differentiates Irregular Warfare from other forms of warfare is its emphasis on the use of irregular forces and indirect methods and means to subvert, attrit and exhaust an enemy, or render him irrelevant rather than to defeat him through direct, conventional military confrontation.

Unlike conventional warfare, which focuses on defeating an adversary's military forces or seizing key physical terrain, the focus of Irregular Warfare is on eroding an enemy's power, influence, and will to exercise political authority over an indigenous population.

[20] Olson, Statement of Vice Admiral Eric T. Olson, September 27, 2006.

[21] IW JOC 9/07, p. 5.

Common characteristics of irregular wars include protraction, intertwining of military and non-military methods, participation by individuals and groups not belonging to the regular armed forces or police of any state, and efforts to gain control of or influence the host population. Irregular Warfare operations may occur as part of traditional warfare or independently.[22]

[I]n most instances the lion's share of the burden in terms of irregular warfare is not uniquely military. It is other. It is information. It's diplomacy. It's the other elements of national power.[23]

Or, as the September 2006 draft of the IW JOC put it:

In either case [of offensive or defensive IW], the ultimate goal of any IW campaign is to promote friendly political authority and influence over, and the support of, the host population while eroding enemies' control, influence, and support.[24]

As noted above, a revised version of the September 2006 draft IW JOC, released in January 2007, acknowledged many of the conceptual and other difficulties associated with IW as an organizing principle.[25] Another difference with the earlier draft was that IW JOC 1/07 argued (as did its successor, IW JOC 9/07) that "[i]nsurgency and counterinsurgency are at the core of IW."[26] Also important is that these two documents noted that counterterrorism operations can be subcomponents of counterinsurgency operations or can stand alone.[27]

---

[22] Mannon, Statement of Brigadier General Otis G. Mannon, September 27, 2006.

[23] Federal News Service, Comments of Mario Mancuso, Deputy Assistant Secretary of Defense for Special Operations and Combating Terrorism, Hearing of the Terrorism and Unconventional Threats Subcommittee of the House Armed Services Committee, Subject: Irregular Warfare Roadmap, September 27, 2006, September 30, 2006.

[24] See DoD, *Irregular Warfare (IW) Joint Operating Concept (JOC)*, draft, Washington, D.C., September 2006, p. 34.

[25] See IW JOC 1/07, especially, pp. 4–5. IW JOC 9/07 contained the same discussion about the "messy" nature of IW, on p. 6.

[26] See IW JOC 9/07, p. 10.

[27] IW JOC 1/07, p. 8, and IW JOC 9/07, p. 10.

Additional strategy-level detail on counterterrorism operations can be found in NMSP-WOT 2/06.[28] For its part, the NMSP-WOT has a slightly different description of the national strategy for the greater war on terrorism and the military strategic framework than does the *National Strategy for Combating Terrorism*, described earlier (see Figures A.3 and A.4).

As described in Figure A.3, the strategy's "ends" are twofold—to defeat violent extremism as a threat to the American way of life as a free and open society and to create a global environment inhospitable to

Figure A.3
National Strategy for Global War on Terrorism

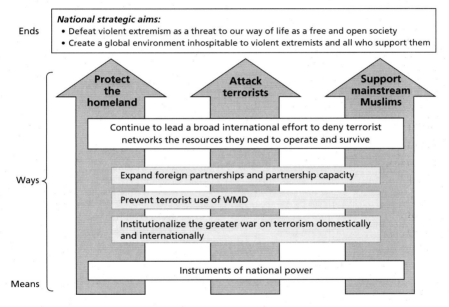

SOURCE: NMSP-WOT 2/06, p. 19.
RAND *MG668-A.3*

---

[28] White House, *National Strategy for Combating Terrorism*, 2006; and Chairman of the Joint Chiefs of Staff, *National Military Strategic Plan for the War on Terrorism*, 2006.

Figure A.4
Military Strategic Framework for Greater War on Terrorism

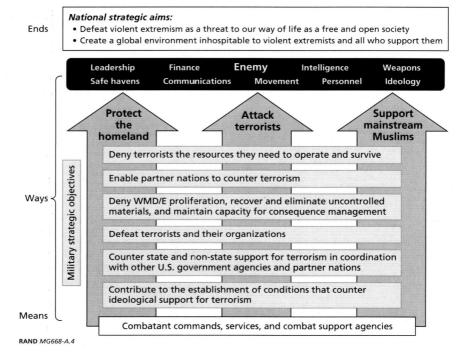

RAND *MG668-A.4*

violent extremists and all who support them—with three "ways" and all instruments of national power supporting these ends.[29]

NMSP-WOT 2/06 similarly describes the military strategy framework for the greater war on terrorism in terms of "ends," "ways," and "means," where the "ways" are a discrete set of military activities that need to be conducted (Figure A.4).

As shown in Figure A.4, NMSP-WOT 2/06 lays out six military strategic objectives:

- deny terrorists the resources they need to operate and survive
- enable partner nations to counter terrorism

---

[29] NMSP-WOT 2/06, p. 19.

- deny WMD/E proliferation, recover and eliminate uncontrolled materials, and maintain capacity for consequence management
- defeat terrorists and their organizations
- counter state and non-state support for terrorism in coordination with other U.S. government agencies and partner nations
- contribute to the establishment of conditions that counter ideological support for terrorism.

It is important to note that NMSP-WOT 2/06 echoes other policy documents in distinguishing between the war on terrorism theater campaign plans of the geographic combatant commands responsible for conducting the greater war on terrorism in theaters like Iraq and Afghanistan, and operations conducted under the SOCOM Global Campaign Plan for the War on Terrorism.[30]

---

[30] NMSP-WOT 2/06, p. 9. Thus, NGIC's analytic requirements theoretically could originate either from SOCOM or from the geographic combatant command or a subordinate command (e.g., Multi-National Corps–Iraq).

Irregular Warfare Analysis Doctrinal References

The following are the doctrinal sources we identified as addressing various aspects of IW analysis:

Air Land Sea Application (ALSA) Center, *Multi-Service Tactics, Techniques, and Procedures for Conducting Peace Operations*, FM 3-07.31, October 2003.

Headquarters, Department of the Army, *Army Special Operations Forces Intelligence*, FM 3-05.102, July 2001. Not releasable to the general public.

————, *Civil Affairs Operations*, FM 41-10, February 2000.

————, *Civil Affairs Operations*, FM 3-05.40, September 2006. Not releasable to the general public.

————, *Civil Affairs Tactics, Techniques, and Procedures*, FM 3-05.401, September 2003.

————, *Counterguerilla Operations*, FM 90-8, August 1986. Not releasable to the general public.

————, *Counterinsurgency*, FM 3-24, December 2006.

————, *Counterinsurgency (Final Draft)*, FM 3-24, June 2006.

————, *Counterintelligence*, FM 34-60, October 1995.

————, *Foreign Internal Defense Tactics, Techniques, and Procedures for Special Forces*, FM 31-20-3, September 1994. Not releasable to the general public.

————, *Human Intelligence Collector Operations*, FM 2-22.3, September 2006.

————, *Intelligence Analysis*, FM 34-3, March 1990.

————, *Intelligence and Electronic Warfare Support to Low-Intensity Conflict Operations,* FM 34-7, May 1993.

————, *Intelligence Preparation of the Battlefield*, FM 34-130, July 1994.

————, *Intelligence Support to Operations in the Urban Environment*, FMI 2-91.4, June 2005. Not releasable to the general public.

————, *Open Source Intelligence*, FMI 2-22.9, December 2006. Not releasable to the general public.

————, *Operations in a Low-Intensity Conflict*, FM 7-98, October 1992.

————, *Police Intelligence Operations*, FM 3-19.50, July 2006.

————, *Psychological Operations Tactics, Techniques, and Procedures*, FM 3-05-30, December 2003. Not releasable to the general public.

————, *Reconnaissance Squadron*, FM 3-20.96, September 2006. Not releasable to the general public.

————, *Special Forces Group Intelligence Operations*, FM 3-05.232, February 2005. Not releasable to the general public.

————, *Urban Operations*, FM 3-06, October 2006.

Joint Chiefs of Staff, *Joint Tactics, Techniques, and Procedures for Foreign Internal Defense (FID)*, JP 3-07.1, 30 April 2004.

U.S. Army Intelligence Center, *Intelligence Support to Stability Operations and Support Operations*, ST 2-91.1, August 2004. Not releasable to the general public.

References

Bueno de Mesquita, Bruce, "The Methodical Study of Politics," paper, October 30, 2002. As of October 29, 2007:
www.yale.edu/probmeth/Bueno_De_Mesquita.doc

Chairman of the Joint Chiefs of Staff, *National Military Strategic Plan for the War on Terrorism*, Washington, D.C., February 1, 2006.

DeNardo, James, *Power in Numbers: The Political Strategy of Protest and Rebellion*, Princeton: Princeton University Press, 1985.

DoD—*See* U.S. Department of Defense

Federal News Service, Comments of Mario Mancuso, Deputy Assistant Secretary of Defense for Special Operations and Combating Terrorism, Hearing of the Terrorism and Unconventional Threats Subcommittee of the House Armed Services Committee, Subject: Irregular Warfare Roadmap, September 27, 2006, September 30, 2006.

Harbom, Lotta, Stina Hogbladh, and Peter Wallensteen, "Armed Conflict and Peace Agreements," *Journal of Peace Research*, Vol. 43, No. 5, 2006.

Headquarters, Department of the Army, *Counterinsurgency*, FM 3-24, Washington, D.C., December 2006.

————, *Full Spectrum Operations*, initial draft, FM 3-0, Washington, D.C., June 21, 2006.

————, *Intelligence*, FM 2-0, Washington, D.C., May 2004.

————, *Intelligence Preparation of the Battlefield*, FM 34-130, Washington, D.C., July 1994.

————, *The Operations Process*, FMI 5-0.1, Washington, D.C., March 2006.

————, *Urban Operations*, FM 3-06, Washington, D.C., October 2006.

Hoffman, Frank G., "Small Wars Revisited: The United States and Nontraditional Wars," *The Journal of Strategic Studies*, Vol. 28, No. 6, December 2005, pp. 913–940.

IW JOC 1/07—*See* Department of Defense, *Irregular Warfare (IW) Joint Operating Concept (JOC)*, Washington, D.C., January 2007

IW JOC 9/07— *See* Department of Defense, *Irregular Warfare (IW) Joint Operating Concept (JOC)*, Version 1.0, Washington, D.C., September 2007

Joint Chiefs of Staff, "Irregular Warfare (IW) Execution Roadmap," unclassified briefing, undated.

Joint Staff, "Joint Capability Areas Taxonomy Tier 1 & Tier 2 with the Initial Draft of Joint Force Projection," briefing, post 24 August 2006 JROC, Washington, D.C., August 2006.

—————, "Proposed Joint Capability Areas Tier 1 and Supporting Tier 2 Lexicon (Mar 06 refinement effort results)," Washington, D.C., March 2006.

Joint Staff J-7/JETCD [Joint Experimentation, Transformation, and Concepts Division], "CJCS JCA Progress Report; SECDEF [Secretary of Defense] Action Memo Tasks; 1-Year Update, 24 August 2006," briefing, Washington, D.C., August 2006.

Kilcullen, David, "Countering Global Insurgency," *The Journal of Strategic Studies*, Vol. 28, No. 4, August 2005, pp. 597–617.

—————, "Countering Global Insurgency," Version 2.2 (earlier of two versions of same paper), *Small Wars Journal,* online edition, November 30, 2004. As of January 2007:
http://www.smallwarsjournal.com/documents/kilcullen.pdf

—————, "Counterinsurgency *Redux*," *Small Wars Journal,* online edition, undated. As of October 29, 2007:
http://www.smallwarsjournal.com/documents/kilcullen1.pdf

—————, "Twenty-Eight Articles: Fundamentals of Company-Level Counterinsurgency," *Iosphere*, Summer 2006, pp. 29–35.

Larson, Eric V., and Bogdan Savych, *Misfortunes of War: Press and Public Reactions to Civilian Deaths in Wartime*, MG-441-AF, Santa Monica, Calif.: RAND Corporation, 2007. As of October 18, 2007:
http://www.rand.org/pubs/monographs/MG441/

Larson, Eric V., Richard E. Darilek, Daniel Gibran, Brian Nichiporuk, Amy Richardson, Lowell H. Schwartz, and Cathryn Quantic Thurston, *Foundations of Effective Influence Operations*, MG-654-A, Santa Monica, Calif.: RAND Corporation, forthcoming.

Larson, Eric V., Richard E. Darilek, Dalia Dassa Kaye, Forrest E. Morgan, Brian Nichiporuk, Diana Dunham-Scott, Cathryn Quantic Thurston, and Kristin J. Leuschner, *Understanding Commanders' Information Needs for Influence Operations*, MG-656-A, Santa Monica, Calif.: RAND Corporation, forthcoming.

Mancuso, Mario, "Irregular Warfare Roadmap," *Special Operations Technology*, online edition. As of January 2007:
http://www.special-operations-technology.com/article.cfm?DocID=1699

———, Statement by Mr. Mario Mancuso, Deputy Assistant Secretary of Defense for Special Operations and Combating Terrorism, Before the 109th Congress, Committee on Armed Services, Subcommittee on Terrorism, Unconventional Threats and Capabilities, United States House of Representatives, September 27, 2006.

Mannon, Otis G., Statement of Brigadier General Otis G. Mannon, U.S. Air Force, Deputy Director, Special Operations, J-3, Joint Staff, Before the 109th Congress Committee on Armed Services, Subcommittee on Terrorism, Unconventional Threats and Capabilities, United States House of Representatives, September 27, 2006.

National Ground Intelligence Center, Web site home page. As of December 2006: http://avenue.org/ngic/index.shtml

NMSP-WOT 2/06—*See* Chairman of the Joint Chiefs of Staff, *National Military Strategic Plan for the War on Terrorism*, Washington, D.C., February 1, 2006

O'Connell, Robert, and John S. White, "NGIC: Penetrating the Fog of War," *Military Intelligence Professional Bulletin*, April–June 2002, pp. 14–18.

Olson, Eric T., Statement of Vice Admiral Eric T. Olson, Deputy Commander, United States Special Operations Command, Before the House Armed Services Committee, Subcommittee on Terrorism, Unconventional Threats and Capabilities, On Irregular Warfare, September 27, 2006.

Packer, George, "Knowing the Enemy," *The New Yorker*, December 18, 2006. As of January 2007:
http://www.newyorker.com

Ray, James Lee, and Bruce Russett, "The Future as Arbiter of Theoretical Controversies: Predictions, Explanations and the End of the Cold War," *British Journal of Political Science*, Vol. 26, 1996, pp. 441–470.

Sherman, Jason, "DoD Plans for Life After the QDR," InsideDefense.com NewsStand, January 12, 2006. As of January 2007:
http://www.military.com/features/0,15240,85093,00.html

———, "New Blueprint for Irregular Warfare," InsideDefense.com NewsStand, May 16, 2006. As of January 2007:
http://www.military.com/features/0,15240,97301,00.html

Sprenger, Sebastian, "DOD, State Dept. Eye Joint 'Hub,'" *Inside the Pentagon*, November 16, 2006. As of October 17, 2007:
http://www.ocnus.net/cgi-bin/exec/view.cgi?archive=105&num=26786

Szayna, Thomas S., ed., *Identifying Potential Ethnic Conflict: Application of a Process Model*, MR-1188-A, Santa Monica, Calif.: RAND Corporation, 2000. As of October 30, 2007:
http://www.rand.org/pubs/monograph_reports/MR1188/

Tarrow, Sidney, *Power in Movement: Social Movements and Contentious Politics*, Cambridge: Cambridge University Press, 1998.

Tilly, Charles, *From Mobilization to Revolution*, New York: McGraw-Hill, 1978.

——————, *Politics of Collective Violence*, Cambridge: Cambridge University Press, 2003.

Under Secretary of Defense for Policy, "Military Support for Stability, Security, Transition, and Reconstruction (SSTR) Operations," DODD 3000.05, November 28, 2005.

United States Air Force, *Irregular Warfare*, AFDD 2-3, Washington, D.C., August 1, 2007.

U.S. Army Combined Arms Doctrine Directorate, "The Continuum of Operations and Stability Operations," briefing, Ft. Leavenworth, Kan.: U.S. Army Combined Arms Center, 2006.

U.S. Department of Defense, *Department of Defense Dictionary of Military and Associated Terms*, JP 1-02, Washington, D.C., April 12, 2001 (as amended through April 14, 2006).

——————, *Doctrine for the Armed Forces of the United States*, Washington, D.C., JP 1, Revision, Final Coordination, October 27, 2006.

——————, *Irregular Warfare (IW) Joint Operating Concept (JOC)*, draft, Washington, D.C., September 2006.

——————, *Irregular Warfare (IW) Joint Operating Concept (JOC)*, Washington, D.C., January 2007.

——————, *Irregular Warfare (IW) Joint Operating Concept (JOC)*, Version 1.0, Washington, D.C., June 2007.

——————, *Irregular Warfare (IW) Joint Operating Concept (JOC)*, Version 1.0, Washington, D.C., September 2007.

——————, "Memorandum for Correspondents," Memorandum No. 046-M, March 2, 1995. As of January 2007:
http://www.defenselink.mil

——————, *National Defense Strategy*, Washington, D.C., June 2008.

——————, *National Defense Strategy of the United States of America*, Washington, D.C., March 2005.

—————, *National Military Strategy of the United States of America: A Strategy for Today; A Vision for Tomorrow*, Washington, D.C., March 2005.

—————, *Quadrennial Defense Review Report*, Washington, D.C., February 2006.

U.S. Joint Forces Command Joint Warfighting Center, *Irregular Warfare Special Study*, Washington, D.C., August 4, 2006.

U.S. Marine Corps, "Small Wars Center of Excellence," Web page, 2007. As of October 17, 2007:
http://www.smallwars.quantico.usmc.mil/

U.S. Marine Corps Combat Development Command, *Tentative Manual for Countering Irregular Threats: An Updated Approach to Counterinsurgency Operations*, Quantico, Va., June 7, 2006.

U.S. Marine Corps Combat Development Command and U.S. Special Operations Command Center for Knowledge and Futures, *Multi-Service Concept for Irregular Warfare*, Version 2.0, August 2, 2006.

White House, *National Strategy for Combating Terrorism*, Washington, D.C., September 2006.

White, Josh, "Gates Sees Terrorism Remaining Enemy No. 1; New Defense Strategy Shifts Focus from Conventional Warfare," *The Washington Post*, July 31, 2008, p. A1.

Wynne, Michael W. [Secretary of the Air Force], "State of the Force," remarks to Air Force Association's Air and Space Conference and Technology Exposition 2006, Washington, D.C., September 25, 2006. As of January 2007:
http://www.af.mil/library/speeches/speech.asp?id=275

Yates, Lawrence A., *The U.S. Military's Experience in Stability Operations, 1789–2005*, Global War on Terrorism Occasional Paper 15, Fort Leavenworth, Kan.: Combat Studies Institute Press, 2006. As of October 18, 2007:
http://www.google.com/search?q=Lawrence+A.+Yates+The+U.S.+Military%27s+Experience&ie=utf-8&oe=utf-8&aq=t&rls=org.mozilla:en-US:official&client=firefox-a